LEARNING MATH THROUGH SOCCER:

Grade 3 Workbook Common Core Edition

Dr. Rodney J. Paul

SEE SPORTS, LLC

Publisher: SEE SPORTS, LLC

Learning Math Through Soccer:
Grade 3 Workbook Common Core Edition

Contact Information:
Rodney J. Paul
info@SeeSports.net
www.SeeSports.net

Printed in the United States of America

Paperback ISBN-13: 978-1-7355240-4-7

Learning Math Through Soccer

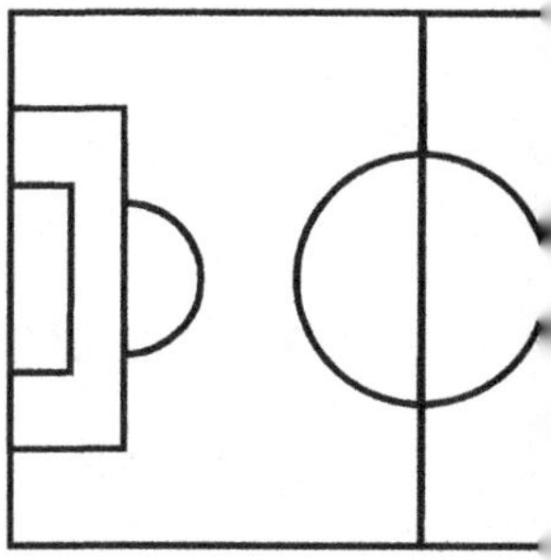

Grade 3 – Common Core Edition

Learning Math Through Soccer uses the excitement of the world's most popular sport to help students understand important math concepts. By connecting math problems to soccer players, stadiums, matches, and training sessions, students can see how math appears in everyday situations.

This book is designed for Grade 3 students and follows Common Core mathematics standards. Each chapter introduces a math concept and connects it to soccer through examples, stories, and practice problems.

Students will explore topics including:

- Multiplication and division
- Word problems and problem solving
- Fractions
- Measurement and time
- Area and perimeter
- Data and graphs
- Basic geometry

Throughout the book, students will encounter famous soccer players, well-known stadiums, and real-world soccer situations while practicing math skills.

The goal is to make learning math **fun, engaging, and meaningful**.

Why Learn Math Through Soccer?

Soccer is played and watched by **billions of people around the world**. The sport includes many situations where math is useful, such as:

- counting players and teams
- measuring field distances
- tracking goals and statistics
- dividing players into practice groups
- analyzing game results

By connecting math to soccer, students can better understand how math is used in the real world.

How to Use This Book

Each chapter includes several sections designed to help students understand and practice math concepts.

Introduction

Explains the math idea and how it connects to soccer.

Key Definitions

Important math vocabulary used in the chapter.

Soccer Connection

Shows how the math concept appears in soccer practices, matches, or statistics.

Did You Know?

Interesting soccer facts about players, stadiums, or tournaments.

Worked Examples

Step-by-step examples that demonstrate how to solve problems.

Practice Problems

Ten problems that allow students to practice the concept.

Challenge Problem

A more advanced problem that encourages deeper thinking.

About the Author

Dr. Rodney J. Paul is Professor, Director, and Department Head of Sport Analytics at Syracuse University's Falk College of Sport. A sports economist with more than 100 peer-reviewed publications, he has spent over twenty years studying betting markets, market efficiency, and gambling behavior. He directs one of the leading undergraduate sport analytics programs in the United States and writes publicly about sports, data, and market behavior through his website, www.SeeSports.net.

This book is part of a series of teaching math through sports for common core, other individual state curricula, and internationally. Additional educational materials related to teaching sport analytics to high school students, specialized academic content, and online versions (app, website) will be available soon.

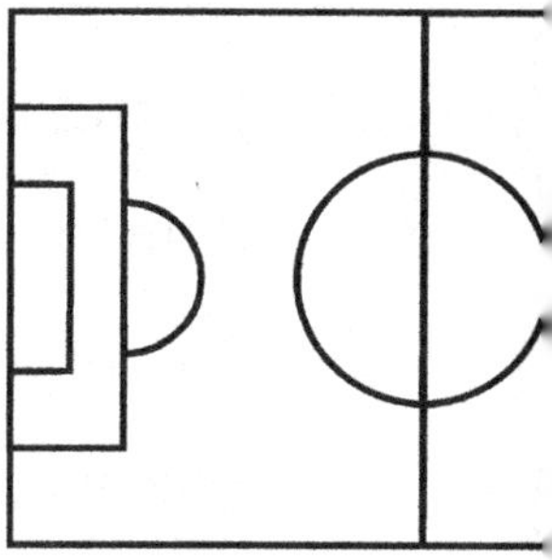

Table of Contents

1

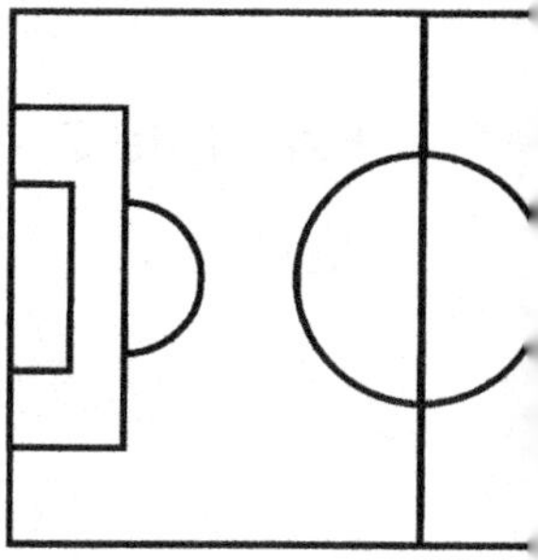

Equal Groups in Soccer Training

Common Core Standard: 3.OA.A.1

Introduction

Soccer coaches often organize players into **equal groups** during practice. For example, a coach might divide players into small groups to work on passing, shooting, or dribbling drills. Learning how to count equal groups helps us understand **multiplication**, which allows us to quickly find totals when groups are the same size.

Key Definitions

Equal Groups: Groups that contain the same number of items.

Multiplication: A way to find the total number of items when equal groups are combined.

Soccer Connection

During soccer training sessions, players often practice in equal groups. At youth academies like **FC Barcelona's La Masia**, coaches organize players into small groups so everyone gets many touches on the ball and improves their skills.

Did You Know?

Lionel Messi began playing organized soccer at age four in Rosario, Argentina. As a young player he trained in small groups that focused on quick passing and ball control—skills that helped him become one of the greatest soccer players in history.

Worked Examples

Example 1

A coach places **3 players in each training group.**
There are **4 groups**.

Add the players in each group.

$3 + 3 + 3 + 3 = 12$

Total players practicing = **12**

Example 2

At **Camp Nou**, a youth drill has **5 groups of 2 players**.

Add the players in each group.

2 + 2 + 2 + 2 + 2 = 10

Total players = **10**

Example 3

During a drill led by **Alex Morgan**, there are **6 groups with 3 players in each group**.

3 + 3 + 3 + 3 + 3 + 3 = 18

Total players = **18**

Practice Problems

1. A coach organizes **5 groups of 3 players** for a passing drill.
   How many players are practicing?

2. At **Wembley Stadium**, a youth team forms **4 groups of 4 players** for shooting practice.
   How many players are in the drill?

3. A coach sets up **6 groups of 2 players** for dribbling practice.
   How many players are practicing?

4. During training, **3 groups of 5 players** practice passing.
   How many players are practicing?

5. At a soccer camp, **7 groups of 2 players** work on ball control.
   How many players are practicing?

6. A coach sets up **4 groups of 6 cones** for dribbling drills.
   How many cones are used?

7. At **Allianz Arena**, **5 groups of 4 players** practice penalty kicks.
   How many players are involved?

8. A team organizes **8 groups of 2 players** for short passing drills.
   How many players are practicing?

9. During a youth tournament, **3 groups of 6 players** warm up together.
   How many players are warming up?

10. Sam Kerr leads a drill with **4 groups of 5 players.**

How many players are in the drill?

Answers

1. 15	2. 16	3. 12	4. 15	5. 14	6. 24	7. 20	8. 16	9. 18	10. 20

Challenge Problem

At **Old Trafford**, a soccer camp has **6 groups of 4 players** practicing passing. Later, the coach adds **2 more groups of 4 players**.

How many players are practicing in total?

Answer: **32**

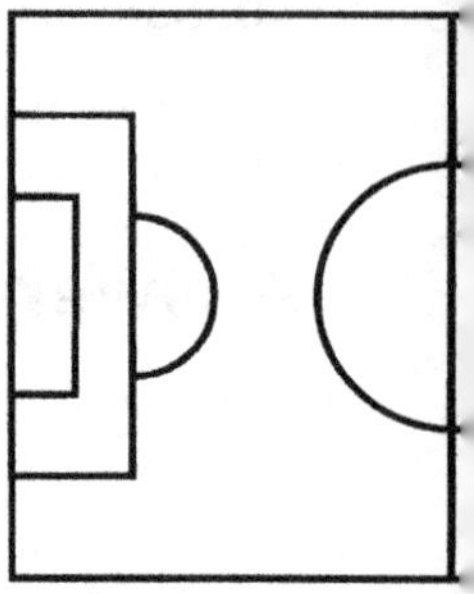

2

Arrays on the Soccer Field

Common Core Standard: 3.OA.A.1

Introduction

In soccer practice, coaches often arrange cones or players in neat rows to organize drills. When objects are placed in rows and columns, they form an **array**. Arrays help us see equal groups and make it easier to understand multiplication.

Key Definitions

Array: An arrangement of objects in rows and columns.

Row: A horizontal line of objects.

Column: A vertical line of objects.

Soccer Connection

During dribbling and passing drills, coaches often place cones in arrays on the field. These organized patterns help players move through drills smoothly and allow coaches to set up practice quickly.

Did You Know?

Kylian Mbappé became one of the youngest stars in World Cup history when he helped France win the **2018 FIFA World Cup**. He was only 19 years old when he scored in the final.

Worked Examples

Example 1

A coach places cones in **3 rows of 4 cones**.

Add the cones in each row.

$4 + 4 + 4 = 12$

Total cones = **12**

Example 2

At **Allianz Arena**, cones are placed in **5 rows of 2 cones** for a passing drill.

2 + 2 + 2 + 2 + 2 = 10

Total cones = **10**

Example 3

A drill has **4 rows of 3 players** practicing short passes.

3 + 3 + 3 + 3 = 12

Total players = **12**

Practice Problems

1. A coach places cones in **4 rows of 5 cones** for a dribbling drill.
   How many cones are used?
2. At **Wembley Stadium**, a drill has **3 rows of 6 players** practicing passing.
   How many players are practicing?
3. A coach sets up **5 rows of 3 cones** for shooting practice.
   How many cones are used?
4. A youth team lines up **2 rows of 7 players** for warmups.
   How many players are there?
5. During practice, **4 rows of 4 cones** are used for dribbling drills.
   How many cones are there?
6. **Trinity Rodman** leads a drill with **3 rows of 5 players.**
   How many players are practicing?
7. At **Camp Nou**, cones are arranged in **6 rows of 2 cones.**
   How many cones are used?
8. A passing drill has **5 rows of 4 players.**
   How many players are practicing?
9. A coach sets up **7 rows of 3 cones** for agility drills.
   How many cones are there?
10. At **Estadio Azteca**, a warmup drill uses **4 rows of 6 players.**

How many players are warming up?

Answers

1. 20	2. 18	3. 15	4. 14	5. 16	6. 15	7. 12	8. 20	9. 21	10. 24

Challenge Problem

At **Maracanã Stadium**, cones are arranged in **5 rows of 6 cones** for a passing drill. The coach later adds **another row of 6 cones**.

How many cones are there in total?

Answer: **36**

3

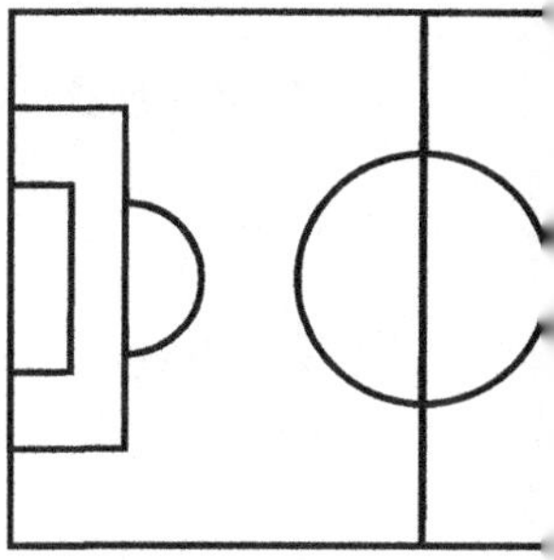

Understanding Multiplication

Common Core Standard: 3.OA.A.1

Introduction

Multiplication helps us quickly find totals when we have **equal groups**. Instead of adding the same number again and again, multiplication gives us a faster way to solve the problem. Soccer training drills often use equal groups, which makes multiplication very useful.

Key Definitions

Multiplication: A math operation used to find the total number in equal groups.

Product: The answer to a multiplication problem.

Soccer Connection

During practice, coaches organize players into equal groups for passing, shooting, and fitness drills. Multiplication helps coaches quickly calculate how many players or pieces of equipment are needed.

Did You Know?

Cristiano Ronaldo is famous for his training habits. As a young player, he spent many hours practicing shooting and dribbling drills in small groups with teammates to improve his skills.

Worked Examples

Example 1

A coach creates **3 groups of 4 players**.

Instead of adding $4 + 4 + 4$, we can multiply.

$3 \times 4 = 12$

Total players = **12**

Example 2

At **Old Trafford**, a drill has **5 groups of 3 players** practicing passing.

$5 \times 3 = 15$

Total players = **15**

Example 3

A training session uses **6 groups of 2 soccer balls**.

$6 \times 2 = 12$

Total soccer balls = **12**

Practice Problems

1. A coach organizes **4 groups of 5 players** for passing drills.
   How many players are practicing?

2. At **Wembley Stadium,** a youth camp forms **3 groups of 7 players.**
   How many players are practicing?

3. A drill uses **6 groups of 3 cones.**
   How many cones are used?

4. A team practices with **5 groups of 4 soccer balls.**
   How many soccer balls are used?

5. **Erling Haaland** leads a drill with 7 **groups of 2 players.**
   How many players are practicing?

6. At **Camp Nou**, a training session has **8 groups of 3 players**.
   How many players are practicing?

7. A coach sets up **4 groups of 6 cones** for dribbling drills.
   How many cones are used?

8. A warmup has **3 groups of 8 players.**
   How many players are warming up?

9. At **Allianz Arena**, **5 groups of 5 players** practice shooting.
   How many players are practicing?

10. A coach organizes **6 groups of 4 players** for passing drills.
    How many players are practicing?

Answers

1. 20	2. 21	3. 18	4. 20	5. 14	6. 24	7. 24	8. 24	9. 25	10. 24

Challenge Problem

At **Estadio Azteca**, a soccer camp organizes **7 groups of 4 players** for passing drills. Later, the coach adds **3 more groups of 4 players**.

How many players are practicing in total?

Answer: **40**

4

Solving Multiplication Word Problems

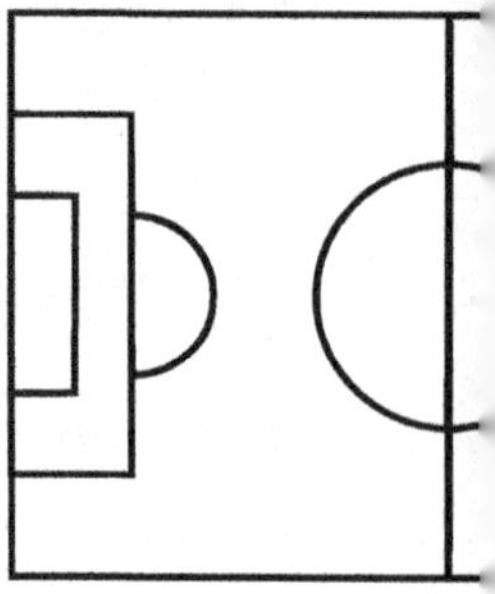

Common Core Standard: 3.OA.A.3

Introduction

Multiplication helps us solve real-world problems involving **equal groups**. In soccer, coaches often organize drills, equipment, and teams in equal groups. Learning how to read and solve word problems helps us understand how multiplication works in everyday situations.

Key Definitions

Word Problem: A math problem written using words that describe a real-life situation.

Equation: A math sentence using numbers and symbols to show a calculation.

Soccer Connection

During soccer practice, coaches must quickly calculate how many players, cones, or soccer balls are needed for drills. Multiplication makes these calculations much easier.

Did You Know?

Alex Morgan helped lead the United States to victory in the **2019 FIFA Women's World Cup**. She scored six goals during the tournament and became one of the most recognizable players in women's soccer.

Worked Examples

Example 1

A coach sets up **4 drills with 5 players in each drill.**

$4 \times 5 = 20$

Total players = **20**

Example 2

At **Wembley Stadium**, a youth tournament has **3 teams with 7 players each**.

$3 \times 7 = 21$

Total players = **21**

Example 3

A training session uses **6 stations with 3 cones at each station**.

$6 \times 3 = 18$

Total cones = **18**

Practice Problems

1. A coach organizes **5 drills with 4 players in each drill**.
   How many players are practicing?
2. At **Old Trafford**, a youth camp forms **3 teams with 8 players each**.
   How many players are there in total?
3. A training drill uses **7 stations with 2 cones at each station**.
   How many cones are used?
4. A team practices with **4 groups of 6 players**.
   How many players are practicing?
5. **Vinícius Jr.** leads a drill with **5 groups of 3 players**.
   How many players are practicing?
6. A coach places **6 rows of 4 cones** on the field for agility drills.
   How many cones are used?
7. At **Camp Nou**, a drill has **3 groups of 9 players**.
   How many players are practicing?
8. A training session uses **8 groups of 2 soccer balls**.
   How many soccer balls are used?
9. At **Allianz Arena**, **5 drills have 5 players each**.
   How many players are practicing?
10. A youth tournament has **4 teams with 7 players each**.
    How many players are playing?

Answers

1. 20	2. 24	3. 14	4. 24	5. 15	6. 24	7. 27	8. 16	9. 25	10. 28

Challenge Problem

At **Maracanã Stadium**, a soccer camp organizes **8 drills with 5 players in each drill**. Later, **2 more drills with 5 players each** join the practice.

How many players are practicing in total?

Answer: **50**

5

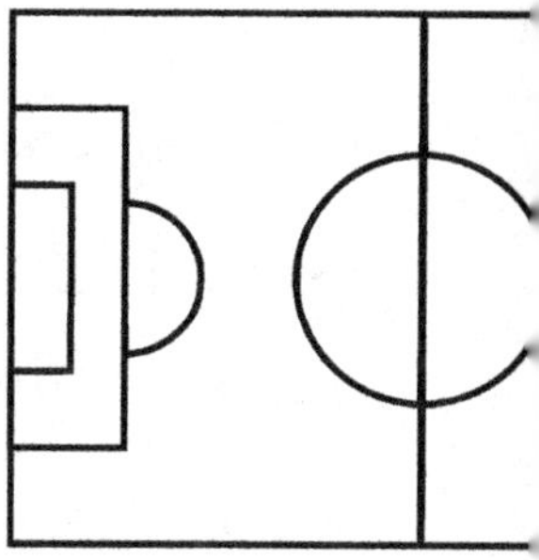

Division and Sharing Soccer Equipment

Common Core Standard: 3.OA.A.2

Introduction

Sometimes in math we need to **share items equally** among groups. This is called **division**. Division helps us find out how many items each group gets when a total amount is split evenly.

Key Definitions

Division: A math operation used to split a total amount into equal groups.

Quotient: The answer to a division problem.

Soccer Connection

During soccer practice, coaches often divide equipment such as soccer balls, cones, or jerseys among players. Division helps ensure that each group receives the same number.

Did You Know?

Marta, one of the greatest women's soccer players ever, has scored more goals in FIFA World Cup tournaments than any other player—male or female. She represented Brazil in multiple World Cups and inspired millions of young players.

Worked Examples

Example 1

A coach has **12 soccer balls** and divides them into **3 equal groups**.

12 ÷ 3 = 4

Each group gets **4 soccer balls**.

Example 2

At **Wembley Stadium**, **18 cones** are divided into **6 equal groups**.

$18 \div 6 = 3$

Each group gets **3 cones**.

Example 3

A team has **20 jerseys** that must be divided among **4 groups**.

$20 \div 4 = 5$

Each group gets **5 jerseys**.

Practice Problems

1. A coach has **16 soccer balls** and divides them into **4 equal groups**.
   How many soccer balls does each group get?

2. At **Camp Nou**, **24 cones** are divided among **6 groups**.
   How many cones does each group get?

3. A team has **15 jerseys** and divides them among **3 groups**.
   How many jerseys does each group get?

4. During practice, **18 soccer balls** are shared among **3 groups**.
   How many soccer balls does each group get?

5. **Sam Kerr** divides **20 cones** among **5 training stations**.
   How many cones are at each station?

6. At **Allianz Arena**, **12 soccer balls** are divided among **4 groups**.
   How many soccer balls does each group receive?

7. A coach shares **21 training bibs** among **3 teams**.
   How many bibs does each team receive?

8. A youth camp divides **24 players** into **6 equal teams**.
   How many players are on each team?

9. At **Old Trafford**, **28 cones** are shared among **4 practice groups**.
   How many cones does each group get?

10. A team shares **18 water bottles** among **3 groups**.
    How many bottles does each group receive?

Answers

1. 4	2. 4	3. 5	4. 6	5. 4	6. 3	7. 7	8. 4	9. 7	10. 6

Challenge Problem

At **Estadio Azteca**, a soccer camp has **36 cones**. The coach divides them equally among **6 training groups**. Later, **another 12 cones** are added and shared equally among the same groups.

How many cones does each group have in total?

Answer: **8**

6

Division Word Problems

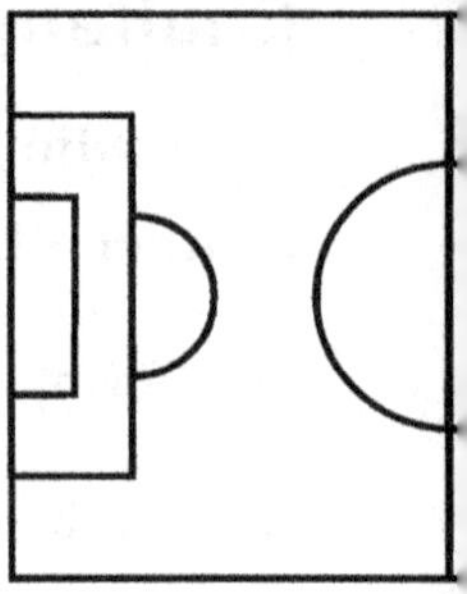

Common Core Standard: 3. OA.A.3

Introduction

Division helps us solve real-world problems when we need to **share a total amount equally**. In soccer, coaches often divide players, equipment, or drills into equal groups so everyone gets a fair chance to participate. Learning how to solve division word problems helps us understand how math works in everyday situations.

Key Definitions

Division: A math operation used to split a total amount into equal groups.

Word Problem: A math problem written using words that describe a real-life situation.

Soccer Connection

During soccer practice, coaches divide players into teams or groups for drills and small games. Division helps coaches quickly determine how many players should be in each group so that practices stay organized and fair.

Did You Know?

Erling Haaland is known for his incredible goal-scoring ability. As a young player in Norway, he trained for many hours each week, often practicing in small groups to improve finishing, passing, and teamwork.

Worked Examples

Example 1

A soccer camp has **20 players**. The coach divides them into **4 equal teams**.

$20 \div 4 = 5$

Each team has **5 players**.

Example 2

At **Wembley Stadium**, a youth training session uses **24 cones** divided among **6 practice stations**.

$24 \div 6 = 4$

Each station gets **4 cones**.

Example 3

A team brings **18 soccer balls** to practice. The coach divides them among **3 groups**.

$18 \div 3 = 6$

Each group gets **6 soccer balls**.

Practice Problems

1. A coach divides **24 players** into **4 equal teams** for a scrimmage.
   How many players are on each team?
2. At **Camp Nou**, **30 cones** are shared among **5 training stations**.
   How many cones does each station receive?
3. A soccer camp has **16 soccer balls** divided among **4 practice groups**.
   How many balls does each group get?
4. **Alex Morgan** divides **21 players** into **3 equal teams**.
   How many players are on each team?
5. At **Allianz Arena**, **36 cones** are shared among **6 drills**.
   How many cones does each drill get?
6. A youth team shares **18 jerseys** among **3 practice groups**.
   How many jerseys are in each group?
7. A training session has **28 players** divided into **4 teams**.
   How many players are on each team?
8. At **Old Trafford**, **32 cones** are divided among **8 stations**.
   How many cones are at each station?
9. A soccer camp divides **27 water bottles** among **3 teams**.
   How many bottles does each team receive?
10. At **Estadio Azteca**, **40 training bibs** are shared among **5 teams**.
    How many bibs does each team get?

Answers

1. 6	2. 6	3. 4	4. 7	5. 6	6. 6	7. 7	8. 4	9. 9	10. 8

Challenge Problem

At **Maracanã Stadium**, a youth soccer camp has **48 players**. The coach divides them into **6 equal teams** for a tournament. Later, **12 more players arrive** and are added evenly to the same teams.

How many players are on each team now?

Answer: **10**

7

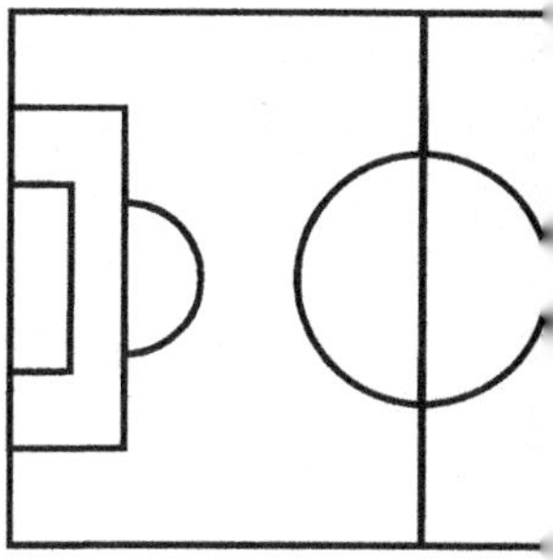

Using Drawings and Equations to Solve Problems

Common Core Standard: 3.OA.A.3

Introduction

When solving math problems, it can be helpful to **draw pictures or write equations** to understand the situation. Drawings can show groups, arrays, or equal parts, while equations use numbers and symbols to represent the same idea. Using both drawings and equations helps us solve problems more clearly and accurately.

Key Definitions

Equation: A math sentence that shows two expressions are equal, using an equal sign (=), and may include operations like addition, subtraction, multiplication, or division.

Model Drawing: A picture or diagram used to help understand and solve a math problem.

Soccer Connection

Soccer coaches often draw diagrams on a whiteboard to explain plays and drills. These diagrams help players understand where to move and how to work together as a team. In math, drawings help us visualize problems in a similar way.

Did You Know?

Pep Guardiola, one of the most successful soccer coaches in the world, is known for drawing detailed diagrams to explain strategies to his players. His teams, including **Manchester City**, often practice using visual diagrams to understand passing patterns and positioning.

Worked Examples

Example 1

A coach organizes **3 groups of 4 players**.

We can draw three groups with four players in each group.

$4 + 4 + 4 = 12$

Equation:

$3 \times 4 = 12$

Total players = **12**

Example 2

At **Wembley Stadium**, a drill has **5 groups of 2 soccer balls.**

$2 + 2 + 2 + 2 + 2 = 10$

Equation:

$5 \times 2 = 10$

Total soccer balls = **10**

Example 3

A team has **18 players** divided into **3 equal groups**.

Equation:

$18 \div 3 = 6$

Each group has **6 players**.

Practice Problems

1. A coach organizes **4 groups of 3 players** for a passing drill. Draw or imagine the groups and write an equation.

   How many players are practicing?

2. At **Camp Nou**, a training session has **5 groups of 4 players**. Write an equation and find the total number of players.

3. A drill uses **3 groups of 6 cones.** Write an equation and find the total number of cones.

4. **Kylian Mbappé** leads a drill with **6 groups of 2 players**. How many players are practicing?

5. At **Allianz Arena**, cones are arranged in **4 groups of 5 cones**. How many cones are used?

6. A youth team divides **24 players** into **4 equal groups**. How many players are in each group?

7. A coach divides **20 soccer balls** into **5 equal groups**.
   How many soccer balls are in each group?

8. At **Old Trafford**, **3 groups of 8 players** practice shooting.
   How many players are practicing?

9. A drill uses **7 groups of 3 cones**.
   How many cones are used?

10. At **Estadio Azteca**, **4 groups of 7 players** warm up before a match.
    How many players are warming up?

Answers

1. 12	2. 20	3. 18	4. 12	5. 20	6. 6	7. 4	8. 24	9. 21	10. 28

Challenge Problem

At **Wembley Stadium**, a soccer camp sets up **5 groups of 6 players** for passing drills. Later, the coach divides the same players into **3 equal teams** for a scrimmage.

How many players are on each team?

Answer: **10**

8

Multiplication and Division Relationships

Common Core Standard: 3.OA.B.5

Introduction

Multiplication and division are closely connected. If we know a multiplication fact, we can often use it to solve a related division problem. Understanding how these operations work together helps us solve math problems more efficiently.

Key Definitions

Multiplication Fact: A multiplication equation such as $4 \times 5 = 20$.

Division Fact: A division equation such as $20 \div 5 = 4$.

Soccer Connection

During soccer training, coaches often organize players and equipment into equal groups. Multiplication can help find the total number of players or cones, while division can help determine how many players belong in each group.

Did You Know?

Megan Rapinoe helped lead the United States to victory in the **2019 FIFA Women's World Cup**. She scored six goals during the tournament and won the Golden Boot as the tournament's top scorer.

Worked Examples

Example 1

Multiplication fact:

$4 \times 5 = 20$

Related division facts:

$20 \div 5 = 4$

$20 \div 4 = 5$

These equations all describe the same relationship.

Example 2

At **Camp Nou**, a drill has **3 groups of 6 players**.

Multiplication:

$3 \times 6 = 18$

Division:

$18 \div 3 = 6$

$18 \div 6 = 3$

Example 3

A coach places **5 cones in each of 4 rows**.

Multiplication:

$4 \times 5 = 20$

Division:

$20 \div 4 = 5$

$20 \div 5 = 4$

Practice Problems

1. A coach organizes **4 groups of 6 players** for a passing drill.
   Find the total number of players using multiplication.
   Then write a related division fact.

2. At **Wembley Stadium**, a drill has **5 groups of 4 cones**.
   Find the total number of cones and write a related division equation.

3. A team forms **3 groups of 8 players**.
   Find the total number of players and a related division equation.

4. **Kylian Mbappé** leads a drill with **6 groups of 3 players**.
   Find the total number of players and a related division fact.

5. At **Allianz Arena**, cones are arranged in **4 groups of 7 cones**.
   Find the total number of cones and write a division equation.

6. A coach organizes **2 groups of 9 players**.
   Find the total number of players and write a related division equation.

7. A training session has **5 groups of 6 soccer balls**.

Find the total number of soccer balls and a related division equation.

8. At **Old Trafford**, a warmup has **3 groups of 9 players**.
   Find the total number of players and a related division equation.

9. A drill uses **8 groups of 3 cones**.
   Find the total number of cones and a related division equation.

10. At **Estadio Azteca**, a coach organizes **4 groups of 8 players**.
    Find the total number of players and write a related division equation.

Answers

1. 24; 24 ÷ 6 = 4 or 24 ÷ 4 = 6	2. 20; 20 ÷ 4 = 5 or 20 ÷ 5 = 4	3. 24; 24 ÷ 8 = 3 or 24 ÷ 3 = 8
4. 18; 18 ÷ 3 = 6 or 18 ÷ 6 = 3	5. 28; 28 ÷ 7 = 4 or 28 ÷ 4 = 7	6. 18; 18 ÷ 2 = 9 or 18 ÷ 9 = 2
7. 30; 30 ÷ 6 = 5 or 30 ÷ 5 = 6	8. 27; 27 ÷ 9 = 3 or 27 ÷ 3 = 9	9. 24; 24 ÷ 3 = 8 or 24 ÷ 8 = 3
10. 32; 32 ÷ 8 = 4 or 32 ÷ 4 = 8		

Challenge Problem

At **Maracanã Stadium**, a youth camp organizes **6 groups of 5 players** for passing drills.

11. How many players are practicing in total?
12. If those players are later divided into **5 equal teams**, how many players are on each team?

Answers:

1. 30	2. 6

9

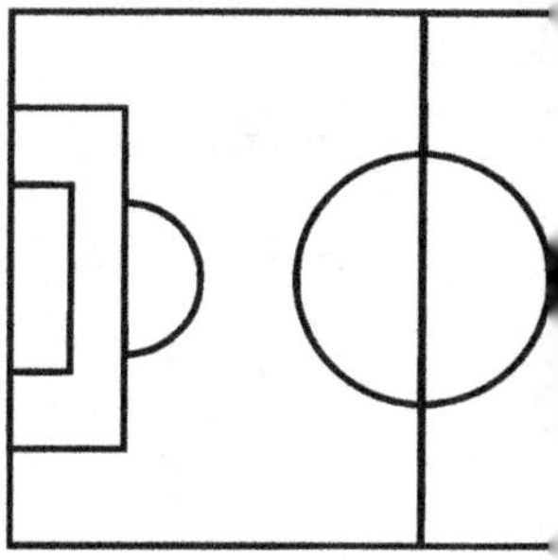

Multiplication Strategies and Patterns

Common Core Standard: 3.OA.D.9

Introduction

Multiplication often follows patterns that make problems easier to solve. By looking for patterns in numbers, we can find answers more quickly and understand how multiplication works. Recognizing these patterns helps us solve problems more efficiently in math and in real-life situations.

Key Definitions

Pattern: A sequence of numbers or objects that follows a predictable rule.

Multiplication Strategy: A method used to solve multiplication problems more easily.

Soccer Connection

Soccer practices often follow patterns. For example, players may repeat drills several times, or teams may rotate through stations in an organized sequence. Just like these patterns in training, multiplication patterns help us understand how numbers grow.

Did You Know?

Pelé, one of the greatest soccer players in history, scored over **1,000 career goals** in official and unofficial matches. His incredible scoring ability made him a global soccer legend.

Worked Examples

Example 1

Multiples of 5 follow a pattern.

$5 \times 1 = 5$

$5 \times 2 = 10$

$5 \times 3 = 15$

$5 \times 4 = 20$

The products increase by **5 each time**.

Example 2

Multiples of 3 also form a pattern.

$3 \times 1 = 3$

$3 \times 2 = 6$

$3 \times 3 = 9$

$3 \times 4 = 12$

Each product increases by **3**.

Example 3

A coach sets up **4 groups of 3 cones.**

$4 \times 3 = 12$

If another group is added:

$5 \times 3 = 15$

The total increases by **3 cones**.

Practice Problems

1. A coach sets up **3 groups of 5 cones** for a passing drill.
   How many cones are used?

2. At **Wembley Stadium**, **4 groups of 5 players** practice shooting.
   How many players are practicing?

3. A drill uses **5 groups of 5 soccer balls**.
   How many soccer balls are used?

4. **Lionel Messi** leads a drill with **6 groups of 5 players**.
   How many players are practicing?

5. A coach organizes **3 groups of 7 players** for scrimmages.
   How many players are practicing?

6. At **Camp Nou**, **4 groups of 7 cones** are used for dribbling drills.
   How many cones are used?

7. A warmup has **5 groups of 6 players**.
   How many players are warming up?

8. At **Allianz Arena**, a coach sets up 7 **groups of 4 cones**.

How many cones are used?

9. A youth camp organizes **8 groups of 3 players** for passing drills.
   How many players are practicing?

10. At **Old Trafford**, **6 groups of 6 players** practice finishing drills.
    How many players are practicing?

Answers

1. 15	2. 20	3. 25	4. 30	5. 21	6. 28	7. 30	8. 28	9. 24	10. 36

Challenge Problem

At **Estadio Azteca**, a coach organizes drills using **groups of 5 players**.

1. If there are **6 groups**, how many players are practicing?
2. If the coach adds **2 more groups**, how many players are practicing now?

Answers:

1. 30	2. 40

10

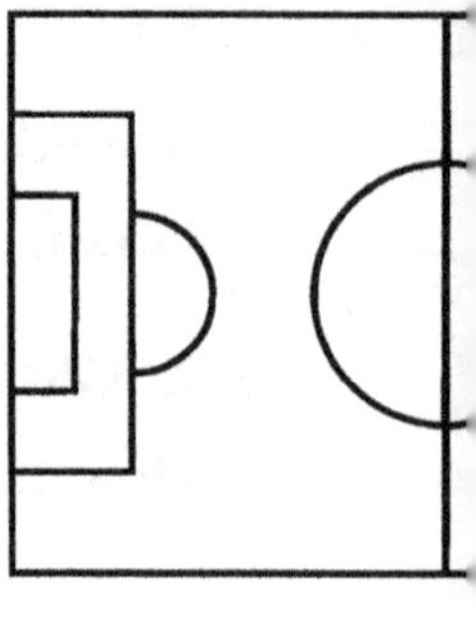

Quarter 1 Soccer Math Challenge

Common Core Standards: 3.OA.A.1, 3.OA.A.2, 3.OA.A.3, 3.OA.B.5, 3.OA.D.9

Introduction

During the first part of this book, we learned how multiplication and division help us understand **equal groups, arrays, and real-world problems**. These skills are used often in soccer training when coaches organize players, equipment, and drills. In this chapter, you will review what you have learned and apply these skills to new soccer situations.

Key Definitions

Multiplication: A way to find the total number in equal groups.

Division: A way to split a total amount into equal groups.

Array: An arrangement of objects in rows and columns.

Soccer Connection

Soccer practices often include many drills happening at the same time. Coaches must organize players, cones, and soccer balls into equal groups so that every player can practice. Multiplication and division help coaches plan training sessions quickly and efficiently.

Did You Know?

The **FIFA World Cup** is the largest soccer tournament in the world. Teams from around the globe compete every four years for the championship, and millions of fans watch the matches.

Worked Examples

Example 1

A coach organizes **4 groups of 6 players** for passing drills.

$4 \times 6 = 24$

Total players = **24**

Example 2

At **Camp Nou**, **24 players** are divided into **6 equal teams**.

$24 \div 6 = 4$

Each team has **4 players**.

Example 3

A drill has **3 rows of 7 cones**.

$7 + 7 + 7 = 21$

Total cones = **21**

Practice Problems

1. A coach organizes **5 groups of 4 players** for a passing drill.
   How many players are practicing?

2. At **Wembley Stadium**, **24 players** are divided into **4 equal teams**.
   How many players are on each team?

3. A drill uses **6 groups of 3 cones**.
   How many cones are used?

4. **Cristiano Ronaldo** leads a drill with **7 groups of 2 players**.
   How many players are practicing?

5. A coach places cones in **4 rows of 6 cones**.
   How many cones are used?

6. At **Allianz Arena**, **30 soccer balls** are divided among **5 training groups**.
   How many balls does each group receive?

7. A youth camp has **3 groups of 8 players** practicing passing drills.
   How many players are practicing?

8. At **Old Trafford**, **18 cones** are divided among **3 stations**.
   How many cones are at each station?

9. A coach organizes **6 groups of 5 players** for a scrimmage.
   How many players are practicing?

10. At **Estadio Azteca**, cones are arranged in **5 rows of 4 cones**.

How many cones are used?

Answers

1. 20	2. 6	3. 18	4. 14	5. 24	6. 6	7. 24	8. 6	9. 30	10. 20

Challenge Problem

At **Maracanã Stadium**, a youth soccer camp organizes **6 groups of 5 players** for a tournament warmup. Later, the coach divides all the players into **5 equal teams** for a scrimmage.

How many players are on each team?

Answer: **6**

11

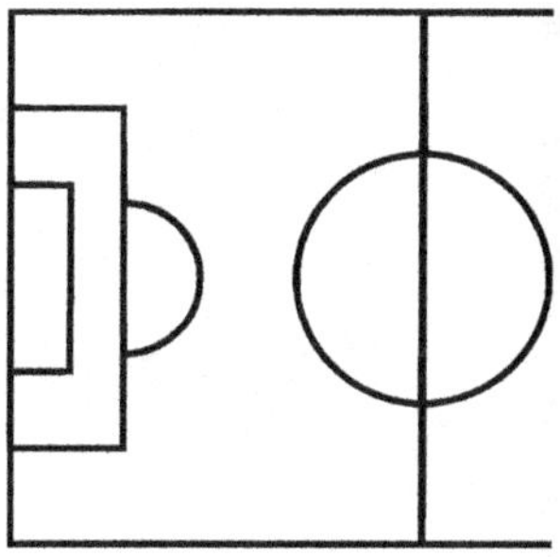

Multiplication Facts: 2s, 5s, and 10s

Common Core Standard: 3.OA.C.7

Introduction

Multiplication facts help us quickly find totals when numbers repeat in equal groups. Some of the easiest facts to learn are the **2s, 5s, and 10s** because they follow clear patterns. These facts are very useful when counting players, cones, or soccer balls during practice.

Key Definitions

Multiplication Fact: A basic multiplication equation such as $5 \times 4 = 20$.

Multiple: The result of multiplying a number by another number.

Soccer Connection

Soccer coaches often organize players and equipment in groups of two, five, or ten. For example, a coach might place cones in groups of five or count players by tens when organizing a large training session.

Did You Know?

Cristiano Ronaldo has scored hundreds of goals in professional soccer. His dedication to training and repetition has helped him become one of the top scorers in soccer history.

Worked Examples

Example 1

A coach sets up **4 groups of 2 cones**.

$4 \times 2 = 8$

Total cones = **8**

Example 2

At **Wembley Stadium**, a drill uses **5 groups of 5 players**.

$5 \times 5 = 25$

Total players = **25**

Example 3

A training session has **3 groups of 10 soccer balls.**

$3 \times 10 = 30$

Total soccer balls = **30**

Practice Problems

1. A coach organizes **6 groups of 2 players** for passing drills.
   How many players are practicing?
2. At **Camp Nou**, **4 groups of 5 cones** are used for dribbling drills.
   How many cones are used?
3. A team practices with **3 groups of 10 soccer balls.**
   How many soccer balls are used?
4. **Kylian Mbappé** leads a drill with **5 groups of 2 players.**
   How many players are practicing?
5. A coach sets up **7 groups of 2 cones** for agility drills.
   How many cones are used?
6. At **Allianz Arena**, **6 groups of 5 players** practice shooting.
   How many players are practicing?
7. A youth camp organizes **8 groups of 2 players** for passing drills.
   How many players are practicing?
8. At **Old Trafford**, **5 groups of 10 cones** are placed on the field.
   How many cones are used?
9. A coach organizes **4 groups of 5 players** for a scrimmage.
   How many players are practicing?
10. At **Estadio Azteca**, **9 groups of 2 players** warm up before a match.
    How many players are warming up?

Answers

1. 12	2. 20	3. 30	4. 10	5. 14	6. 30	7. 16	8. 50	9. 20	10. 18

Challenge Problem

At **Maracanã Stadium**, a soccer camp sets up **8 groups of 5 players** for passing drills. Later, the coach adds **2 more groups of 5 players**.

How many players are practicing in total?

Answer: **50**

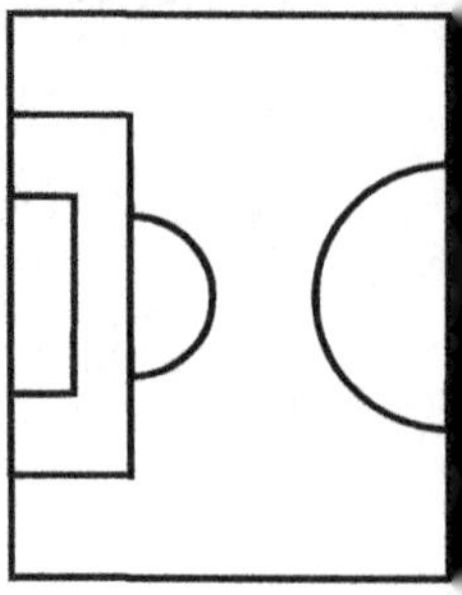

12

Multiplication Facts: 3s and 4s

Common Core Standard: 3.OA.C.7

Introduction

Multiplication facts help us quickly calculate totals when numbers repeat in equal groups. The **3s and 4s multiplication facts** are especially useful because many soccer drills involve groups of three or four players. Learning these facts helps players and coaches organize practice quickly.

Key Definitions

Multiplication Fact: A basic multiplication equation such as 3 × 4 = 12.

Product: The answer to a multiplication equation.

Soccer Connection

Many soccer drills involve groups of three or four players working together. For example, small passing drills often use three players passing the ball in a triangle, while four players may practice moving and passing in a square formation.

Did You Know?

Zinedine Zidane, one of the greatest midfielders in soccer history, was known for his incredible control and passing ability. He helped France win the **1998 FIFA World Cup** and later became a successful coach for **Real Madrid**.

Worked Examples

Example 1

A coach organizes **4 groups of 3 players** for passing drills.

4 × 3 = 12

Total players = **12**

Example 2

At **Wembley Stadium**, cones are arranged in **3 rows of 4 cones**.

$3 \times 4 = 12$

Total cones = **12**

Example 3

A drill uses **5 groups of 4 players**.

$5 \times 4 = 20$

Total players = **20**

Practice Problems

1. A coach sets up **6 groups of 3 players** for a passing drill.
   How many players are practicing?

2. At **Camp Nou**, cones are arranged in **4 rows of 3 cones**.
   How many cones are used?

3. A drill has **5 groups of 3 players** practicing shooting.
   How many players are practicing?

4. **Erling Haaland** leads a drill with 7 **groups of 3 players**.
   How many players are practicing?

5. A coach organizes **4 groups of 4 players** for a scrimmage.
   How many players are practicing?

6. At **Allianz Arena**, cones are arranged in **6 rows of 4 cones**.
   How many cones are used?

7. A youth camp organizes **3 groups of 8 players** for passing drills.
   How many players are practicing?

8. At **Old Trafford**, **5 groups of 4 players** practice shooting drills.
   How many players are practicing?

9. A coach places 7 **groups of 4 cones** on the field.
   How many cones are used?

10. At **Estadio Azteca**, **8 groups of 3 players** warm up before a match.
    How many players are warming up?

Answers

1. 18	2. 12	3. 15	4. 21	5. 16	6. 24	7. 24	8. 20	9. 28	10. 24

Challenge Problem

At **Maracanã Stadium**, a coach organizes **6 groups of 4 players** for passing drills. Later, the coach adds **3 more groups of 4 players**.

How many players are practicing in total?

Answer: **36**

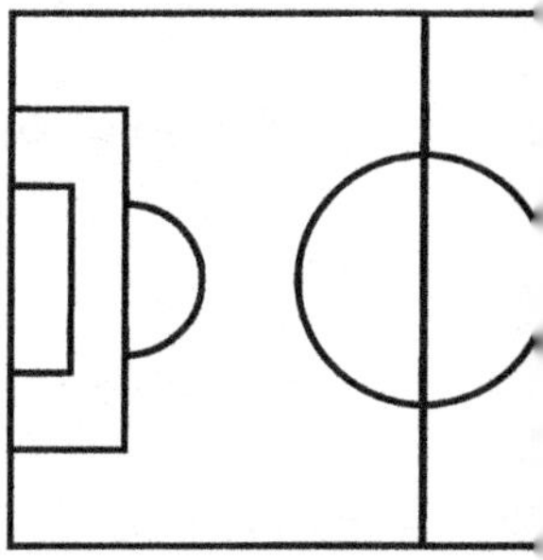

13

Multiplication Facts: 6s and 7s

Common Core Standard: 3.OA.C.7

Introduction

Multiplication facts with **6s and 7s** help us solve many real-world problems. These facts appear often in soccer situations when coaches organize drills, teams, or equipment. Learning these multiplication facts helps players and coaches calculate totals quickly during practice.

Key Definitions

Multiplication Fact: A basic multiplication equation such as 6 × 4 = 24.

Product: The answer to a multiplication problem.

Soccer Connection

Many soccer drills use groups of six or seven players. Coaches might divide players into small teams for scrimmages or place cones in groups to organize training stations. Multiplication helps coaches quickly count the total number of players or pieces of equipment.

Did You Know?

Kylian Mbappé became the second teenager in history to score in a FIFA World Cup Final when France won the **2018 World Cup**. His speed and scoring ability have made him one of the most exciting players in modern soccer.

Worked Examples

Example 1

A coach organizes **4 groups of 6 players** for a passing drill.

4 × 6 = 24

Total players = **24**

Example 2

At **Wembley Stadium**, cones are placed in **3 rows of 7 cones**.

$3 \times 7 = 21$

Total cones = **21**

Example 3

A drill has **5 groups of 6 players** practicing shooting.

$5 \times 6 = 30$

Total players = **30**

Practice Problems

1. A coach organizes **6 groups of 6 players** for a passing drill.
   How many players are practicing?
2. At **Camp Nou**, cones are arranged in **4 rows of 7 cones**.
   How many cones are used?
3. A drill has **5 groups of 7 players** practicing shooting.
   How many players are practicing?
4. **Lionel Messi** leads a drill with **3 groups of 6 players**.
   How many players are practicing?
5. A coach organizes 7 **groups of 6 cones** for agility drills.
   How many cones are used?
6. At **Allianz Arena**, cones are arranged in **6 rows of 7 cones**.
   How many cones are used?
7. A youth camp organizes **4 groups of 6 players** for passing drills.
   How many players are practicing?
8. At **Old Trafford**, **5 groups of 6 players** practice finishing drills.
   How many players are practicing?
9. A coach sets up **8 groups of 7 cones** for dribbling drills.
   How many cones are used?
10. At **Estadio Azteca**, 7 **groups of 6 players** warm up before a match.
    How many players are warming up?

Answers

1. 36	2. 28	3. 35	4. 18	5. 42	6. 42	7. 24	8. 30	9. 56	10. 42

Challenge Problem

At **Maracanã Stadium**, a soccer camp organizes **6 groups of 7 players** for passing drills. Later, the coach adds **4 more groups of 7 players**.

How many players are practicing in total?

Answer: **70**

14

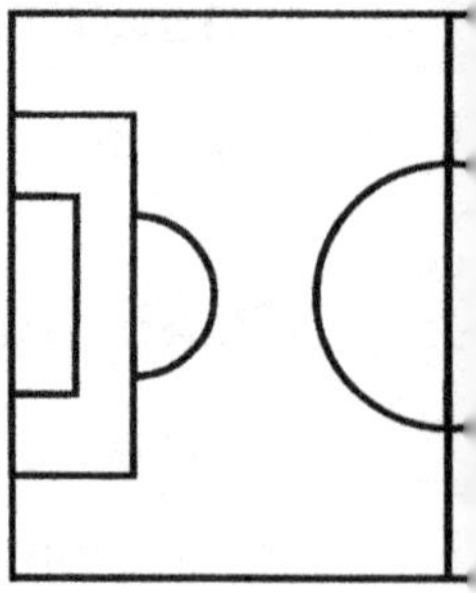

Multiplication Facts: 8s and 9s

Common Core Standard: 3.OA.C.7

Introduction

Multiplication facts with **8s and 9s** can seem harder at first, but they become easier when we practice them regularly. These facts appear often in real-world situations, including soccer training when coaches organize players, drills, and equipment. Learning these multiplication facts helps us solve problems more quickly and confidently.

Key Definitions

Multiplication Fact: A basic multiplication equation such as 8 × 3 = 24.

Product: The answer to a multiplication equation.

Soccer Connection

Soccer teams often practice in groups of eight or nine players during drills or scrimmages. Coaches may also set up multiple training stations, each with the same number of cones or soccer balls. Multiplication helps coaches quickly determine the total number needed.

Did You Know?

Luka Modrić, the Croatian midfielder, won the **2018 Ballon d'Or**, an award given to the best soccer player in the world. He was known for his incredible passing, vision, and leadership on the field.

Worked Examples

Example 1

A coach organizes **3 groups of 8 players** for passing drills.

3 × 8 = 24

Total players = **24**

Example 2

At **Camp Nou**, cones are arranged in **4 rows of 9 cones.**

$4 \times 9 = 36$

Total cones = **36**

Example 3

A drill uses **5 groups of 8 soccer balls.**

$5 \times 8 = 40$

Total soccer balls = **40**

Practice Problems

1. A coach organizes **6 groups of 8 players** for a passing drill.
   How many players are practicing?

2. At **Wembley Stadium**, cones are arranged in **5 rows of 9 cones.**
   How many cones are used?

3. A drill has **7 groups of 8 players** practicing shooting.
   How many players are practicing?

4. **Vinícius Jr.** leads a drill with **4 groups of 9 players.**
   How many players are practicing?

5. A coach organizes **8 groups of 8 cones** for agility drills.
   How many cones are used?

6. At **Allianz Arena**, cones are arranged in **3 rows of 9 cones.**
   How many cones are used?

7. A youth camp organizes **5 groups of 8 players** for passing drills.
   How many players are practicing?

8. At **Old Trafford**, **6 groups of 9 players** practice finishing drills.
   How many players are practicing?

9. A coach sets up **7 groups of 8 cones** for dribbling drills.
   How many cones are used?

10. At **Estadio Azteca**, **8 groups of 9 players** warm up before a match.
    How many players are warming up?

Answers

1. 48	2. 45	3. 56	4. 36	5. 64	6. 27	7. 40	8. 54	9. 56	10. 72

Challenge Problem

At **Maracanã Stadium**, a soccer camp organizes **7 groups of 8 players** for a passing drill. Later, the coach adds **3 more groups of 8 players**.

How many players are practicing in total?

Answer: **80**

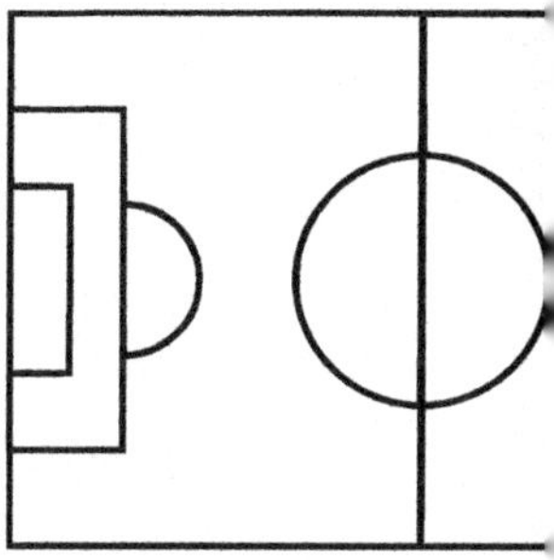

15

Two-Step Soccer Word Problems

Common Core Standard: 3.OA.D.8

Introduction

Some math problems require **more than one step** to solve. These are called **two-step problems**. In soccer, coaches often plan drills that involve several actions, such as organizing groups and then combining players for another activity. Learning how to solve problems step by step helps us find the correct answer.

Key Definitions

Two-Step Problem: A problem that requires two calculations to solve.

Equation: A math sentence using numbers and symbols to represent a problem.

Soccer Connection

During soccer practice, coaches may first divide players into groups and then combine those groups for scrimmages or drills. Solving two-step problems helps coaches plan these activities quickly and accurately.

Did You Know?

Trinity Rodman, one of the rising stars of the United States Women's National Team, became one of the youngest players drafted in the **National Women's Soccer League (NWSL)**. She is known for her speed, creativity, and goal-scoring ability.

Worked Examples

Example 1

A coach organizes **4 groups of 5 players** for passing drills.

First step:

$4 \times 5 = 20$ players

Then the coach adds **5 more players**.

Second step:

20 + 5 = 25

Total players = **25**

Example 2

At **Camp Nou**, cones are arranged in **3 rows of 6 cones**.

First step:

3 × 6 = 18 cones

Then the coach adds **4 more cones**.

Second step:

18 + 4 = 22

Total cones = **22**

Example 3

A youth camp has **5 groups of 4 players** practicing.

First step:

5 × 4 = 20 players

Then **8 more players arrive**.

Second step:

20 + 8 = 28

Total players = **28**

Practice Problems

1. A coach organizes **5 groups of 4 players** for a drill. Later, **6 more players join**. How many players are practicing in total?
2. At **Wembley Stadium**, a drill uses **4 rows of 6 cones**. Then the coach adds **8 more cones**. How many cones are used in total?
3. A youth camp has **3 groups of 7 players** practicing passing. Later, **5 more players join**. How many players are practicing?
4. **Kylian Mbappé** leads a drill with **6 groups of 3 players**. Then **4 more players join the drill**. How many players are practicing in total?

5. A coach sets up **5 groups of 5 cones** for dribbling drills. Then **5 more cones are added.**
   How many cones are used in total?

6. At **Allianz Arena**, a training session has **4 groups of 8 players**. Later, **8 more players arrive**.
   How many players are practicing?

7. A coach organizes **3 groups of 9 players**. Then **6 more players join**.
   How many players are practicing?

8. At **Old Trafford**, cones are placed in **6 rows of 4 cones**. Then **12 more cones are added**.
   How many cones are used?

9. A drill uses **5 groups of 6 players**. Then **10 more players join**.
   How many players are practicing?

10. At **Estadio Azteca**, a youth camp has **4 groups of 7 players** practicing shooting. Later, **7 more players arrive**.
    How many players are practicing?

Answers

1. 26	2. 32	3. 26	4. 22	5. 30	6. 40	7. 33	8. 36	9. 40	10. 35

Challenge Problem

At **Maracanã Stadium**, a soccer camp organizes **6 groups of 6 players** for drills. Later, **12 more players arrive** and join the practice. The coach then divides all the players into **6 equal teams** for a scrimmage.

How many players are on each team?

Answer: **8**

16

Using Parentheses in Soccer Calculations

Common Core Standard: 3.OA.D.8

Introduction

Sometimes math problems include **more than one operation**, such as multiplication and addition. Parentheses help show **which part of the problem should be solved first**. Learning to use parentheses helps us solve multi-step problems in the correct order.

Key Definitions

Parentheses: Symbols ( ) used in math to show which calculation should be done first.

Order of Operations: The rule that tells us the order in which we solve parts of a math problem.

Soccer Connection

During soccer training, coaches often organize drills in steps. For example, they might first combine several groups of players and then add more players to the activity. Parentheses help show the correct order for solving these types of problems.

Did You Know?

Andrés Iniesta, one of Spain's greatest midfielders, scored the winning goal in the **2010 FIFA World Cup Final**. His goal helped Spain win its first World Cup championship.

Worked Examples

Example 1

A drill has **3 groups of 4 players**, and then **2 more players join**.

$(3 \times 4) + 2$

First solve inside the parentheses:

$3 \times 4 = 12$

Then add:

$12 + 2 = 14$

Total players = **14**

Example 2

At **Camp Nou**, cones are arranged in **4 groups of 5 cones**, and then **5 more cones are added**.

$(4 \times 5) + 5$

$4 \times 5 = 20$

$20 + 5 = 25$

Total cones = **25**

Example 3

A soccer camp has **2 groups of 8 players**, and then **6 more players arrive**.

$(2 \times 8) + 6$

$2 \times 8 = 16$

$16 + 6 = 22$

Total players = **22**

Practice Problems

1. A coach organizes **4 groups of 5 players**, and then **5 more players join**.
   How many players are practicing?

2. At **Wembley Stadium**, cones are arranged in **3 groups of 6 cones**, and then **6 more cones are added**.
   How many cones are used?

3. A drill has **5 groups of 4 players**, and then **8 more players join**.
   How many players are practicing?

4. **Lionel Messi** leads a drill with **6 groups of 3 players**, and then **6 more players join**.
   How many players are practicing?

5. A coach sets up **4 groups of 7 cones**, and then **7 more cones are added**.
   How many cones are used?

6. At **Allianz Arena**, a training session has **3 groups of 8 players**, and then **4 more players arrive**.
   How many players are practicing?

7. A youth camp has **5 groups of 6 players**, and then **6 more players join**.

How many players are practicing?

8. At **Old Trafford**, cones are arranged in **4 groups of 9 cones**, and then **9 more cones are added**.
   How many cones are used?

9. A coach organizes **2 groups of 10 players**, and then **5 more players join**.
   How many players are practicing?

10. At **Estadio Azteca**, a drill has **3 groups of 7 players**, and then **7 more players join**.
    How many players are practicing?

Answers

1. 25	2. 24	3. 28	4. 24	5. 35	6. 28	7. 36	8. 45	9. 25	10. 28

Challenge Problem

At **Maracanã Stadium**, a soccer camp organizes **5 groups of 6 players** for passing drills. Later, **12 more players arrive** and join the practice.

Write and solve the equation:

$(5 \times 6) + 12$

$5 \times 6 = 30$

$30 + 12 =$ **42**

Answer: **42 players**

17

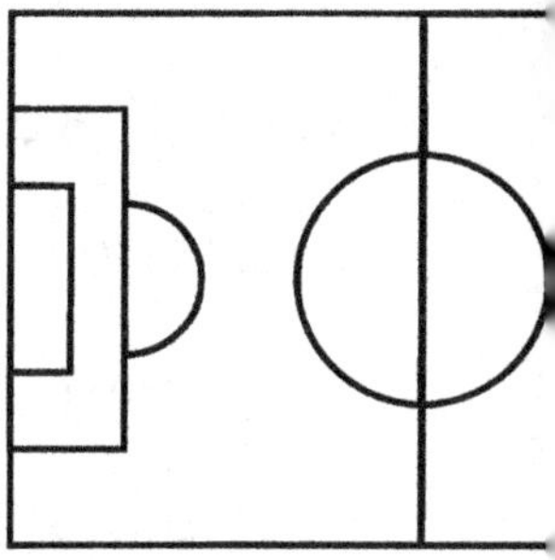

Identifying Arithmetic Patterns

Common Core Standard: 3.OA.D.9

Introduction

In math, numbers often follow **patterns**. A pattern happens when numbers increase or decrease in a predictable way. Recognizing patterns helps us understand multiplication, solve problems faster, and make predictions about what comes next.

Key Definitions

Pattern: A sequence of numbers that follows a rule.

Arithmetic Pattern: A pattern where numbers increase or decrease by the same amount each time.

Soccer Connection

Patterns appear often in soccer training. Coaches might repeat the same drill several times or increase the number of players or cones in each round of practice. Recognizing patterns helps players stay organized and helps coaches plan effective training sessions.

Did You Know?

Johan Cruyff, one of the most influential players and coaches in soccer history, helped develop the famous **"Total Football"** style of play used by the Netherlands and later by FC Barcelona. His ideas changed how teams move and pass the ball on the field.

Worked Examples

Example 1

A coach counts players in groups of **3**.

3, 6, 9, 12, 15

The pattern increases by **3 each time**.

Example 2

A drill adds **5 cones each round**.

5, 10, 15, 20

The pattern increases by **5 each time**.

Example 3

A team practices in groups of **4 players**.

4, 8, 12, 16

The pattern increases by **4 each time**.

Practice Problems

1. A coach counts players in groups of **3**.
   3, 6, 9, 12, ___
   What number comes next?

2. At **Camp Nou**, cones are added in groups of **5**.
   5, 10, 15, ___
   What number comes next?

3. A passing drill increases by **4 players each round**.
   4, 8, 12, ___
   What number comes next?

4. **Alex Morgan** counts goals scored in groups of **2**.
   2, 4, 6, 8, ___
   What number comes next?

5. A coach adds **6 cones each round.**
   6, 12, 18, ___
   What number comes next?

6. At **Allianz Arena**, players warm up in groups of **7**.
   7, 14, 21, ___
   What number comes next?

7. A training drill increases by **8 cones each round**.
   8, 16, 24, ___
   What number comes next?

8. At **Old Trafford**, players practice penalty kicks in groups of **5**.

5, 10, 15, 20, ___

What number comes next?

9. A youth camp increases the number of players by **9 each round**.

   9, 18, 27, ___

   What number comes next?

10. At **Estadio Azteca**, cones are placed in groups of **4**.

    4, 8, 12, 16, ___

    What number comes next?

Answers

1. 15	2. 20	3. 16	4. 10	5. 24	6. 28	7. 32	8. 25	9. 36	10. 20

Challenge Problem

At **Maracanã Stadium**, a coach adds **6 players to a drill each round**.

Round 1: 6 players
Round 2: 12 players
Round 3: 18 players

If the pattern continues, how many players will there be in **Round 6**?

Answer: **36**

18

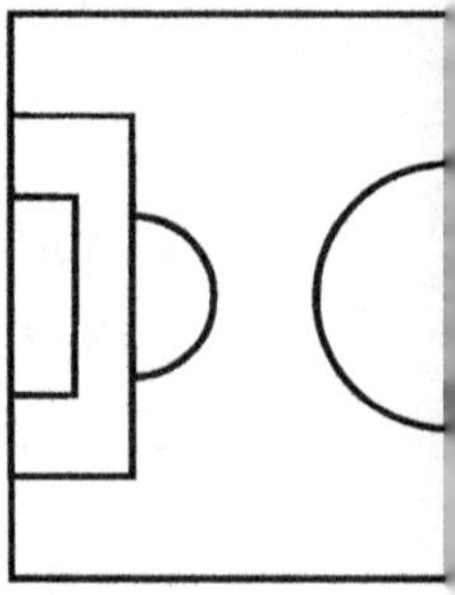

Problem Solving with Soccer Statistics

Common Core Standard: 3.OA.D.8

Introduction

Soccer teams keep track of many statistics during games and tournaments. These statistics include goals scored, assists, passes, and saves. Math helps us understand these numbers and solve problems based on game results and player performance.

Key Definitions

Statistic: A number that describes information about a game or performance.

Total: The sum of several numbers added together.

Soccer Connection

Coaches and analysts often review statistics to understand how a team or player is performing. For example, they might count the total number of goals scored in several games or calculate how many passes a player makes during a match.

Did You Know?

Lionel Messi has won multiple **Ballon d'Or** awards, which are given each year to the best soccer player in the world. His incredible goal scoring and playmaking have helped his teams win many championships.

Worked Examples

Example 1

A player scores **2 goals in each of 4 games**.

$2 \times 4 = 8$

Total goals scored = **8**

Example 2

At **Camp Nou**, a team scores **3 goals in each of 5 matches.**

$3 \times 5 = 15$

Total goals = **15**

Example 3

A goalkeeper makes **4 saves in each of 6 matches.**

$4 \times 6 = 24$

Total saves = **24**

Practice Problems

1. A player scores **3 goals in each of 4 games.**
   How many goals does the player score in total?

2. At **Wembley Stadium**, a team scores **2 goals in each of 6 matches.**
   How many goals are scored in total?

3. A goalkeeper makes **5 saves in each of 4 games.**
   How many saves are made in total?

4. **Kylian Mbappé** scores **2 goals in each of 5 games.**
   How many goals does he score?

5. A team completes **6 passes during each practice drill** and runs the drill **4 times.**
   How many passes are completed?

6. At **Allianz Arena**, a player takes **3 shots in each of 7 games.**
   How many shots are taken?

7. A defender makes **4 tackles in each of 5 games.**
   How many tackles are made?

8. At **Old Trafford**, a team scores **3 goals in each of 6 matches.**
   How many goals are scored?

9. A goalkeeper makes **2 saves in each of 9 games.**
   How many saves are made?

10. At **Estadio Azteca**, a team completes **5 corner kicks in each of 4 matches.**
    How many corner kicks are taken?

Answers

1. 12	2. 12	3. 20	4. 10	5. 24	6. 21	7. 20	8. 18	9. 18	10. 20

Challenge Problem

At **Maracanã Stadium**, a player scores **3 goals in each of 6 matches**. Later in the season, the player scores **4 more goals in another game**.

How many goals has the player scored in total?

First step:

$3 \times 6 = 18$

Second step:

$18 + 4 = \mathbf{22}$

Answer: **22 goals**

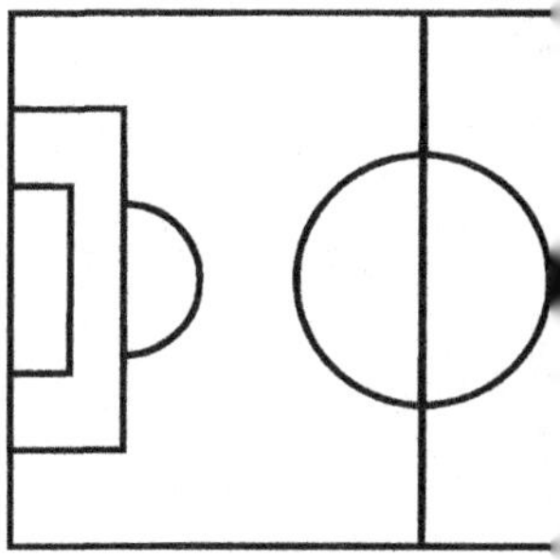

19

Review of Multiplication and Division

Common Core Standards: 3.OA.A.1, 3.OA.A.2, 3.OA.A.3, 3.OA.C.7

Introduction

Multiplication and division help us solve many problems involving **equal groups**. In soccer, these skills are useful when organizing players, dividing equipment, or planning drills. In this chapter, we will review how multiplication and division work together to solve problems.

Key Definitions

Multiplication: A way to find the total number in equal groups.

Division: A way to split a total number into equal groups.

Soccer Connection

During soccer practice, coaches often divide players into teams or set up drills with equal numbers of cones or soccer balls. Multiplication helps find totals, while division helps determine how many items belong in each group.

Did You Know?

Cristiano Ronaldo has played for several famous clubs, including **Manchester United**, **Real Madrid**, and **Juventus**. He is known for his incredible goal-scoring ability and dedication to training.

Worked Examples

Example 1

A coach organizes **4 groups of 6 players**.

$4 \times 6 = 24$

Total players = **24**

Example 2

A team has **24 players** and divides them into **4 equal teams**.

$24 \div 4 = 6$

Each team has **6 players**.

Example 3

A drill has **5 groups of 5 cones**.

$5 \times 5 = 25$

Total cones = **25**

Practice Problems

1. A coach organizes **3 groups of 7 players**.
   How many players are practicing?
2. At **Camp Nou**, **20 cones** are divided into **5 groups**.
   How many cones are in each group?
3. A drill uses **6 groups of 4 players**.
   How many players are practicing?
4. **Erling Haaland** leads a drill with **5 groups of 6 players**.
   How many players are practicing?
5. A coach divides **24 players** into **6 teams**.
   How many players are on each team?
6. At **Allianz Arena**, a drill has **4 rows of 8 cones**.
   How many cones are used?
7. A team divides **18 soccer balls** into **3 groups**.
   How many balls are in each group?
8. At **Old Trafford**, **7 groups of 4 players** practice shooting.
   How many players are practicing?
9. A youth camp divides **30 players** into **5 equal teams**.
   How many players are on each team?
10. At **Estadio Azteca**, **8 groups of 5 players** practice passing.
    How many players are practicing?

Answers

1. 21	2. 4	3. 24	4. 30	5. 4	6. 32	7. 6	8. 28	9. 6	10. 40

Challenge Problem

At **Maracanã Stadium**, a soccer camp organizes **5 groups of 8 players** for training drills. Later, the coach divides all the players into **4 equal teams** for a scrimmage.

How many players are on each team?

First step:

$5 \times 8 = 40$

Second step:

$40 \div 4 =$ **10**

Answer: **10 players**

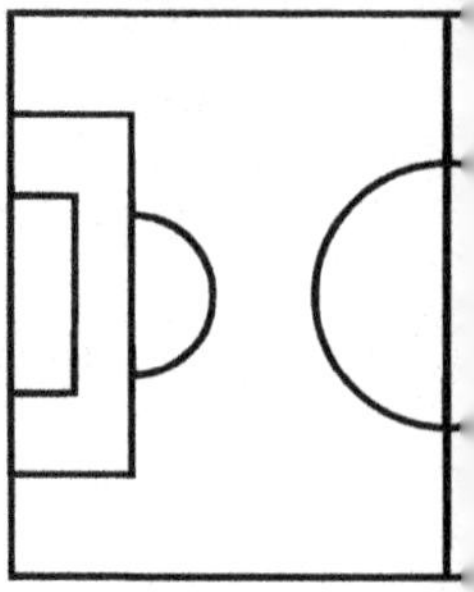

20

Quarter 2 Tournament Challenge

Common Core Standards: 3.OA.A.1, 3.OA.A.2, 3.OA.A.3, 3.OA.C.7, 3.OA.D.8

Introduction

During the second part of this book, we practiced multiplication facts, division, and solving multi-step problems. These skills are useful in many real-world situations, including soccer tournaments where teams, players, and equipment must be organized carefully. In this chapter, you will review what you have learned and apply it to new soccer challenges.

Key Definitions

Multiplication: A way to find the total number in equal groups.

Division: A way to split a total number into equal groups.

Two-Step Problem: A problem that requires more than one calculation.

Soccer Connection

Soccer tournaments require careful planning. Coaches must divide players into teams, organize drills, and keep track of statistics during matches. Multiplication and division help make these tasks faster and easier.

Did You Know?

The **UEFA Champions League** is one of the most famous club soccer tournaments in the world. Teams from many European countries compete each year for the championship.

Worked Examples

Example 1

A youth tournament has **4 teams with 7 players on each team**.

$4 \times 7 = 28$

Total players = **28**

Example 2

At **Wembley Stadium**, **30 players** are divided into **5 equal teams**.

$30 \div 5 = 6$

Each team has **6 players**.

Example 3

A drill uses **5 groups of 6 cones**, and then **4 more cones are added**.

$5 \times 6 = 30$

$30 + 4 = 34$

Total cones = **34**

Practice Problems

1. A tournament has **5 teams with 6 players on each team.**
   How many players are playing?
2. At **Camp Nou**, **24 players** are divided into **4 equal teams**.
   How many players are on each team?
3. A drill uses **6 groups of 5 cones**.
   How many cones are used?
4. **Lionel Messi** leads a drill with **4 groups of 8 players**.
   How many players are practicing?
5. A coach divides **36 cones** into **6 equal groups**.
   How many cones are in each group?
6. At **Allianz Arena**, a training session has **3 groups of 9 players**.
   How many players are practicing?
7. A youth camp has **4 groups of 7 players**, and then **7 more players join**.
   How many players are practicing?
8. At **Old Trafford**, **28 players** are divided into **4 teams**.
   How many players are on each team?
9. A drill uses **5 groups of 8 cones**.
   How many cones are used?
10. At **Estadio Azteca**, a coach organizes **6 groups of 6 players**.
    How many players are practicing?

Answers

1. 30	2. 6	3. 30	4. 32	5. 6	6. 27	7. 35	8. 7	9. 40	10. 36

Challenge Problem

At **Maracanã Stadium**, a youth soccer tournament has **6 teams with 7 players on each team**. Later, **14 more players arrive** and join the tournament. The coach divides all the players into **7 equal teams**.

How many players are on each team?

First step:

$6 \times 7 = 42$

Second step:

$42 + 14 = 56$

Third step:

$56 \div 7 = \mathbf{8}$

Answer: **8 players per team**

21

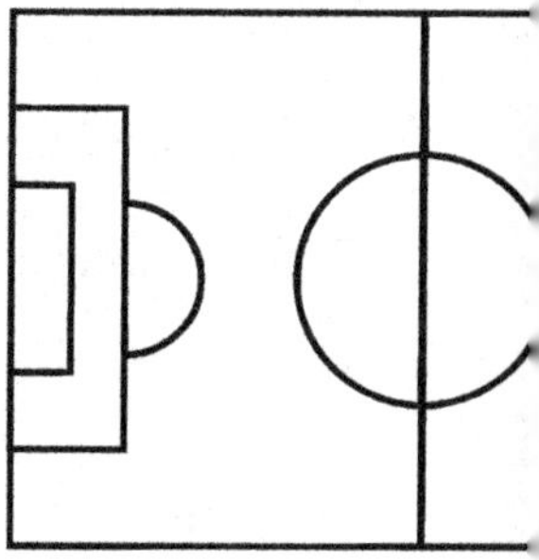

Understanding Fractions of a Soccer Field

Common Core Standard: 3.NF.A.1

Introduction

A **fraction** represents part of a whole. In soccer, we often divide the field into equal sections during practice. Fractions help describe these parts and allow players and coaches to understand how much of the field is being used.

Key Definitions

Fraction: A number that represents part of a whole.

Numerator: The top number of a fraction that shows how many parts we have.

Denominator: The bottom number of a fraction that shows how many equal parts the whole is divided into.

Soccer Connection

During training sessions, coaches often divide the soccer field into equal sections so that multiple drills can take place at the same time. For example, one drill might use half the field while another uses the other half.

Did You Know?

A professional soccer field is usually about **100–110 meters long**. Coaches often divide the field into smaller sections during practice to help players focus on specific skills such as passing or defending.

Worked Examples

Example 1

A coach divides the soccer field into **2 equal parts**.

Each part is:

1/2 of the field.

Example 2

At **Camp Nou**, a coach divides the field into **4 equal sections.**

Each section is:

1/4 of the field.

Example 3

A drill uses **3 of the 4 sections** of the field.

This is:

3/4 of the field.

Practice Problems

1. A soccer field is divided into **2 equal parts.**
   What fraction of the field is each part?

2. At **Wembley Stadium**, the field is divided into **4 equal sections** for practice.
   What fraction of the field is one section?

3. A drill uses **1 of the 3 equal sections** of a practice field.
   What fraction of the field is being used?

4. A team uses **2 of the 4 equal sections** of the field for a passing drill.
   What fraction of the field is being used?

5. A coach divides the field into **5 equal sections.**
   What fraction of the field is one section?

6. At **Allianz Arena,** a drill uses **3 of 5 equal sections** of the field.
   What fraction of the field is being used?

7. A coach divides the field into **6 equal sections** for training drills.
   What fraction of the field is one section?

8. At **Old Trafford**, players practice in **2 of the 6 equal sections** of the field.
   What fraction of the field is being used?

9. A drill uses **4 of the 8 equal sections** of the field.
   What fraction of the field is being used?

10. At **Estadio Azteca**, the field is divided into **10 equal sections.**
    What fraction of the field is one section?

Answers

1. 1/2	2. 1/4	3. 1/3	4. 2/4	5. 1/5	6. 3/5	7. 1/6	8. 2/6	9. 4/8	10. 1/10

Challenge Problem

At **Maracanã Stadium**, the field is divided into **8 equal sections** for practice drills. One drill uses **3 sections**, and another drill uses **2 sections**.

What fraction of the field is being used in total?

3/8 + 2/8 = **5/8**

Answer: **5/8 of the field**

22

Fractions on the Number Line

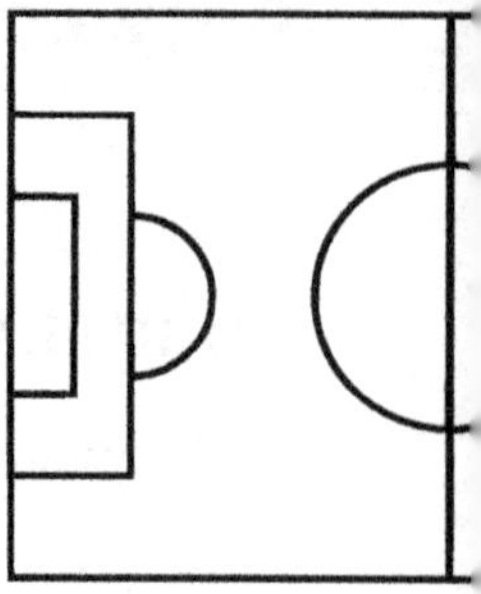

Common Core Standard: 3.NF.A.2

Introduction

Fractions can be shown in different ways. One useful way is by placing fractions on a **number line**. A number line helps us see where fractions belong between whole numbers. This helps us understand how fractions represent equal parts of a whole.

Key Definitions

Number Line: A straight line used to show numbers in order.

Fraction: A number that represents part of a whole.

Soccer Connection

Soccer matches are timed events. A match lasts **90 minutes**, often divided into two halves. Coaches sometimes think about time during a match using fractions, such as half the match or one quarter of the match.

Did You Know?

A professional soccer match has **two halves of 45 minutes each**. This means each half represents **one-half (1/2)** of the entire match.

Worked Examples

Example 1

A number line from **0 to 1** is divided into **2 equal parts**.

The point in the middle represents:

1/2

Example 2

A number line from **0 to 1** is divided into **4 equal parts**.

The points represent:

1/4, 2/4, 3/4

Example 3

A number line from **0 to 1** is divided into **3 equal parts.**

The points represent:

1/3, 2/3

Practice Problems

1. A number line from **0 to 1** is divided into **2 equal parts.**
   What fraction is the middle point?
2. A number line from **0 to 1** is divided into **4 equal parts.**
   What fraction is the first point after 0?
3. A number line from **0 to 1** is divided into **3 equal parts.**
   What fraction is the first point after 0?
4. During a soccer match, **half the match has been played.**
   What fraction of the match has been completed?
5. At **Wembley Stadium**, a match reaches **one quarter of the total time.**
   What fraction of the match has been played?
6. A number line from **0 to 1** is divided into **5 equal parts.**
   What fraction is the second point after 0?
7. A number line from **0 to 1** is divided into **6 equal parts.**
   What fraction is the third point after 0?
8. At **Camp Nou**, a match is **three quarters finished.**
   What fraction of the match has been played?
9. A number line from **0 to 1** is divided into **8 equal parts.**
   What fraction is the fourth point after 0?
10. A number line from **0 to 1** is divided into **10 equal parts.**
    What fraction is the fifth point after 0?

Answers

1. 1/2 | 2. 1/4 | 3. 1/3 | 4. 1/2 | 5. 1/4 | 6. 2/5 | 7. 3/6 | 8. 3/4 | 9. 4/8 | 10. 5/10

Challenge Problem

A soccer match lasts **90 minutes**. At **45 minutes**, halftime begins.

What fraction of the match has been played?

45 / 90 = **1/2**

Answer: **1/2 of the match**

23

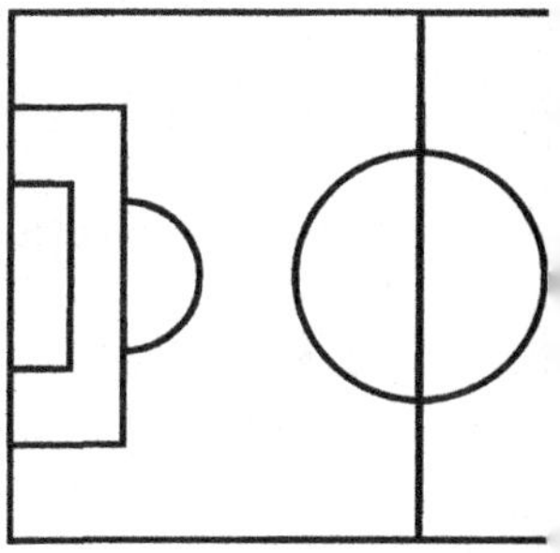

Fractions of a Team

Common Core Standard: 3.NF.A.1

Introduction

Fractions help us describe **parts of a group**. In soccer, teams are often divided into smaller groups during training drills. Fractions help us understand what portion of the team is doing a certain activity.

Key Definitions

Fraction: A number that represents part of a whole or part of a group.

Whole: The total number of items or members in a group.

Soccer Connection

During practice, coaches may divide a team into smaller groups to work on different skills. For example, some players may practice shooting while others practice passing. Fractions help describe how much of the team is doing each activity.

Did You Know?

A standard soccer team has **11 players on the field** during a match. These players include defenders, midfielders, forwards, and a goalkeeper.

Worked Examples

Example 1

A team has **8 players practicing shooting** out of **8 total players**.

Fraction practicing shooting:

8/8

This means the **whole team** is practicing shooting.

Example 2

A team has **4 players practicing passing** out of **8 total players**.

Fraction practicing passing:

4/8

Example 3

A team has **3 players practicing defense** out of **6 players**.

Fraction practicing defense:

3/6

Practice Problems

1. A team has **10 players**, and **5 players practice shooting**.
   What fraction of the team is practicing shooting?

2. At **Wembley Stadium**, a team has **8 players**, and **4 players practice passing**.
   What fraction of the team is practicing passing?

3. A team has **12 players**, and **3 players practice goalkeeping drills**.
   What fraction of the team is practicing goalkeeping?

4. **Lionel Messi** leads a drill where **6 of 12 players practice dribbling**.
   What fraction of the team is practicing dribbling?

5. A coach divides **9 players**, and **3 practice defense**.
   What fraction of the team practices defense?

6. At **Camp Nou**, **4 of 10 players practice shooting**.
   What fraction of the team is practicing shooting?

7. A team has **6 players**, and **2 players practice passing**.
   What fraction of the team practices passing?

8. At **Allianz Arena**, **5 of 15 players practice finishing drills**.
   What fraction of the team is practicing finishing?

9. A team has **8 players**, and **6 players practice attacking drills**.
   What fraction of the team practices attacking?

10. At **Old Trafford**, **3 of 9 players practice defending drills**.
    What fraction of the team practices defending?

Answers

1. 5/10	2. 4/8	3. 3/12	4. 6/12	5. 3/9	6. 4/10	7. 2/6	8. 5/15	9. 6/8	10. 3/9

Challenge Problem

At **Estadio Azteca**, a team has 12 **players practicing.**
4 players practice shooting, **4 practice passing**, and **4 practice defending**.

What fraction of the team practices **each activity**?

Shooting: 4/12
Passing: 4/12
Defending: 4/12

Answer: **Each activity uses 4/12 of the team**

24

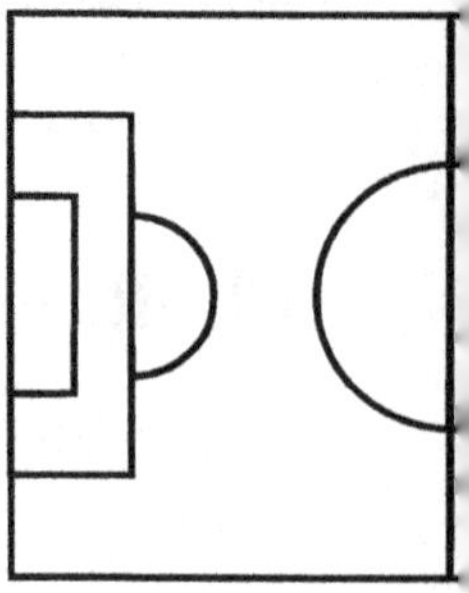

Comparing Fractions

Common Core Standard: 3.NF.A.3

Introduction

Sometimes we need to compare fractions to see **which part is larger or smaller**. Fractions that represent parts of the same whole can be compared using the symbols >, <, **or** =. Understanding how to compare fractions helps us make decisions and solve problems in real-life situations.

Key Definitions

Greater Than (>): A symbol showing that one number is larger than another.

Less Than (<): A symbol showing that one number is smaller than another.

Equal To (=): A symbol showing that two numbers are the same.

Soccer Connection

During practice, coaches may divide a team into different groups for drills. Fractions can help compare how much of the team is working on one skill compared with another skill.

Did You Know?

Cristiano Ronaldo scored more than **800 career goals** in professional matches. His ability to score consistently helped many of his teams win championships.

Worked Examples

Example 1

Compare:

1/2 and 1/4

Half of something is larger than one quarter.

1/2 > 1/4

Example 2

Compare:

3/4 and 2/4

Both fractions have the same denominator.

3/4 > 2/4

Example 3

Compare:

2/6 and 3/6

3/6 is larger because the numerator is larger.

3/6 > 2/6

Practice Problems

1. Compare the fractions:
   1/2 ___ 1/3

2. Compare the fractions:
   3/4 ___ 2/4

3. Compare the fractions:
   2/5 ___ 4/5

4. At **Wembley Stadium**, **3/8 of the team practices shooting** and **2/8 practices passing**.
   Which fraction is larger?

5. Compare:
   5/6 ___ 3/6

6. At **Camp Nou**, **4/10 of players practice passing** and **6/10 practice shooting**.
   Which fraction is larger?

7. Compare:
   1/3 ___ 2/3

8. At **Allianz Arena**, **5/12 of players practice attacking drills** and **4/12 practice defending drills**.
   Which fraction is larger?

9. Compare:
   7/8 ___ 6/8

10. At **Old Trafford**, **3/9 of players practice shooting** and **5/9 practice passing**.
    Which fraction is larger?

Answers

1. 1/2 > 1/3	2. 3/4 > 2/4	3. 2/5 < 4/5	4. 3/8 > 2/8	5. 5/6 > 3/6
6. 6/10 > 4/10	7. 1/3 < 2/3	8. 5/12 > 4/12	9. 7/8 > 6/8	10. 5/9 > 3/9

Challenge Problem

At **Maracanã Stadium**, a team divides practice time between two drills.

3/6 of the time is spent practicing passing.
2/6 of the time is spent practicing shooting.

Which drill takes more time?

3/6 > 2/6

Answer: **Passing practice takes more time.**

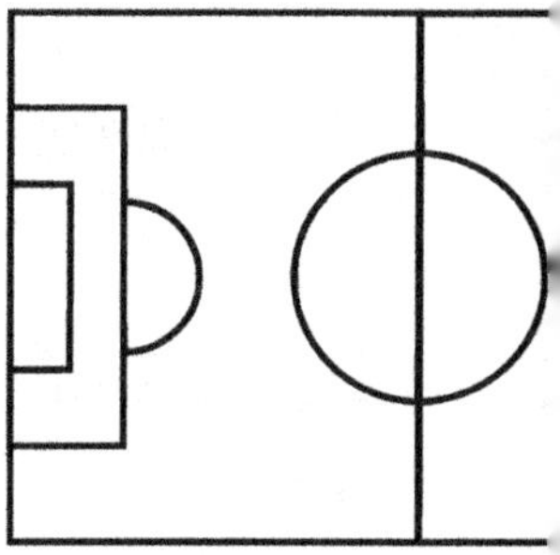

25

Equivalent Fractions with Soccer Examples

Common Core Standard: 3.NF.A.3

Introduction

Sometimes two fractions look different but represent the **same amount**. These are called **equivalent fractions**. Equivalent fractions help us understand that the same part of a whole can be written in different ways.

Key Definitions

Equivalent Fractions: Fractions that represent the same value even though they use different numbers.

Simplify: To write a fraction in a simpler but equal form.

Soccer Connection

During practice, a coach might divide the soccer field into different numbers of sections depending on the drill. Even though the sections may be divided differently, they can represent the same portion of the field.

Did You Know?

The **Maracanã Stadium** in Rio de Janeiro once hosted more than **190,000 spectators** for a soccer match. It remains one of the most famous stadiums in the world.

Worked Examples

Example 1

A field is divided into **2 equal parts**, and **1 part is used**.

1/2

If the same field is divided into **4 equal parts**, the same area might be:

2/4

So:

1/2 = 2/4

Example 2

A team practices on **2 out of 3 sections of the field**.

2/3

If each section is divided again into smaller parts, the same portion might be:

4/6

So:

2/3 = 4/6

Example 3

A team uses **3 out of 4 sections of the field**.

3/4

This is the same as:

6/8

So:

3/4 = 6/8

Practice Problems

1. Are these fractions equivalent?
   1/2 and 2/4

2. Are these fractions equivalent?
   2/3 and 4/6

3. Are these fractions equivalent?
   3/4 and 6/8

4. At **Wembley Stadium**, a drill uses **2/4 of the field**.
   Is this the same as **1/2 of the field**?

5. Are these fractions equivalent?
   3/6 and 1/2

6. At **Camp Nou**, a team practices on **4/8 of the field**.
   Is this the same as **1/2 of the field**?

7. Are these fractions equivalent?
   2/5 and 4/10

8. At **Allianz Arena**, a team practices on **3/6 of the field**.
   Is this equal to **1/2 of the field**?
9. Are these fractions equivalent?
   5/10 and 1/2
10. At **Old Trafford**, a team uses **6/12 of the field**.
    Is this equal to **1/2 of the field**?

Answers

1. Yes	2. Yes	3. Yes	4. Yes	5. Yes	6. Yes	7. Yes	8. Yes	9. Yes	10. Yes

Challenge Problem

At **Estadio Azteca**, a coach divides the soccer field into **8 equal sections**. A drill uses **4 of the sections**.

1. What fraction of the field is used?
2. Write an equivalent fraction using **2 as the denominator**.

Answers:

1. 4/8	2. 1/2

26

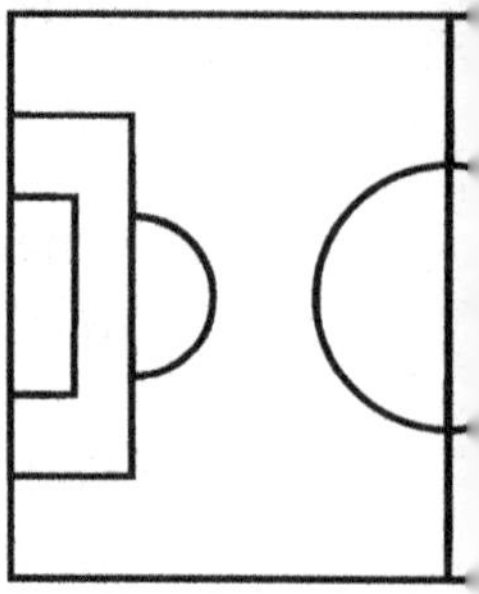

Fractions in Match Time

Common Core Standard: 3.NF.A.1, 3.NF.A.2

Introduction

Fractions can help us understand **time during a soccer match**. A full professional match lasts **90 minutes**, divided into two equal halves. Fractions can describe how much of the match has been played or how much time remains.

Key Definitions

Fraction: A number that represents part of a whole.

Half: One of two equal parts of something.

Soccer Connection

During matches, commentators and coaches often talk about fractions of the game. For example, halftime occurs when **one-half of the match has been played**, and teams may adjust their strategy depending on how much time remains.

Did You Know?

The **FIFA World Cup Final** is one of the most watched sporting events in the world. Hundreds of millions of fans watch the match live around the globe.

Worked Examples

Example 1

A soccer match lasts **90 minutes**.

After **45 minutes**, halftime begins.

45 is **1/2 of 90**, so:
Half the match has been played.

Example 2

If **30 minutes** of a match have passed:

30/90 = 1/3

So **one-third of the match** has been played.

Example 3

If **60 minutes** of a match have passed:

60/90 = 2/3

So **two-thirds of the match** has been played.

Practice Problems

1. A soccer match lasts **90 minutes**.
   After **45 minutes**, what fraction of the match has been played?

2. At **Wembley Stadium, 30 minutes** of a match have passed.
   What fraction of the match has been played?

   If **60 minutes** have passed in a 90-minute match, what fraction of the match has been played?

3. **Lionel Messi** scores a goal after **45 minutes** of a match.
   What fraction of the match has been played?

4. At **Camp Nou**, a goal is scored after **30 minutes**.
   What fraction of the match has been played?

5. A match has **90 minutes**, and **15 minutes** have passed.
   What fraction of the match has been played?

6. At **Allianz Arena, 75 minutes** of a match have passed.
   What fraction of the match has been played?

7. A team scores after **45 minutes**, and the match continues for another **45 minutes**.
   What fraction of the match remains after halftime?

8. At **Old Trafford, 60 minutes** of a match have passed.
   What fraction of the match remains?

9. At **Estadio Azteca, 30 minutes remain in the match**.
   What fraction of the match is left?

Answers

1. 1/2	2. 1/3	3. 2/3	4. 1/2	5. 1/3	6. 1/6	7. 5/6	8. 1/2	9. 1/3	10. 1/3

Challenge Problem

A soccer match lasts **90 minutes**.

A goal is scored after **30 minutes**, and another goal is scored after **60 minutes**.

1. What fraction of the match had been played when the first goal was scored?
2. What fraction of the match had been played when the second goal was scored?

Answers:

1. 1/3	2. 2/3

27

Fractions of Practice Drills

Common Core Standard: 3.NF.A.1

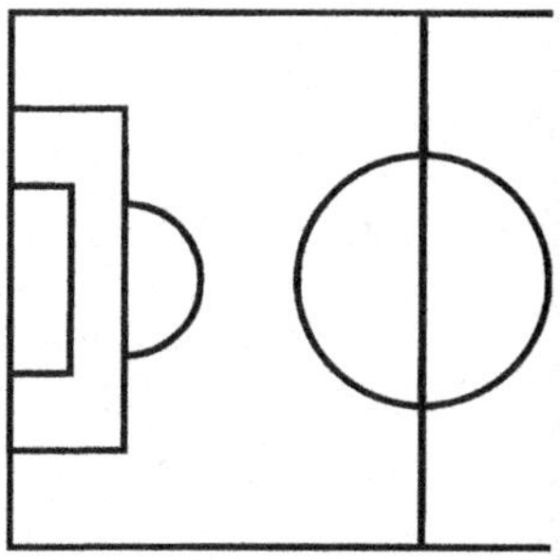

Introduction

Fractions help describe **parts of activities** during soccer practice. Coaches often divide practice into different drills such as passing, shooting, defending, and conditioning. Fractions can help show how much of the practice time is spent on each activity.

Key Definitions

Fraction: A number that represents part of a whole.

Whole: The total amount, such as the entire practice session.

Soccer Connection

During a soccer practice, a coach may divide time into several equal parts. For example, one part of practice might focus on passing drills while another part focuses on shooting or defensive skills.

Did You Know?

Professional soccer teams often train for **several hours each day** during the season. Practices usually include warm-ups, skill drills, tactical training, and scrimmages.

Worked Examples

Example 1

A practice session is divided into **4 equal drills**.

Each drill represents:

1/4 of the practice.

Example 2

A coach spends **2 of the 4 drills practicing passing**.

This fraction is:

2/4

Example 3

A team spends **3 of the 5 drills practicing shooting**.

This fraction is:

3/5

Practice Problems

1. A practice session is divided into **3 equal drills**.
   What fraction of practice is one drill?
2. At **Camp Nou**, a coach divides practice into **4 drills**.
   What fraction of practice is each drill?
3. A team spends **2 of 5 drills practicing passing**.
   What fraction of practice is spent on passing?
4. **Kylian Mbappé** leads a training session with **3 of 6 drills focused on shooting**.
   What fraction of practice is spent on shooting?
5. A team spends **4 of 8 drills practicing defense**.
   What fraction of practice is spent on defense?
6. At **Allianz Arena**, **2 of 4 drills focus on dribbling**.
   What fraction of practice is spent on dribbling?
7. A team spends **3 of 7 drills practicing passing**.
   What fraction of practice is spent on passing?
8. At **Old Trafford**, **5 of 10 drills focus on shooting**.
   What fraction of practice is spent on shooting?
9. A coach divides practice into **6 equal drills**.
   What fraction of practice is one drill?
10. At **Estadio Azteca**, **4 of 6 drills focus on defense**.
    What fraction of practice is spent on defense?

Answers

1. 1/3	2. 1/4	3. 2/5	4. 3/6	5. 4/8	6. 2/4	7. 3/7	8. 5/10	9. 1/6	10. 4/6

Challenge Problem

A soccer practice session is divided into **8 equal drills**.

- **3 drills focus on passing**
- **2 drills focus on shooting**

What fraction of practice is spent on **passing and shooting combined**?

3/8 + 2/8 = **5/8**

Answer: **5/8 of practice** is spent on passing and shooting.

28

Soccer Strategy with Fractions

Common Core Standard: 3.NF.A.3

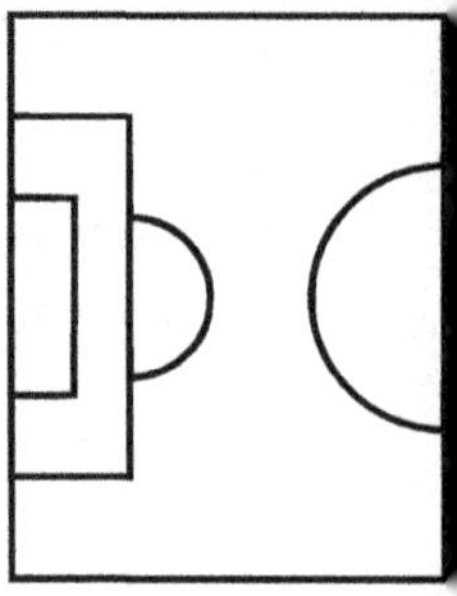

Introduction

Fractions can help us describe **parts of a strategy** during a soccer match. Coaches often divide time, players, or field space into parts to plan how a team will attack or defend. Understanding fractions helps us compare these parts and see which one is larger.

Key Definitions

Fraction: A number that represents part of a whole.

Compare: To decide whether one number is greater than, less than, or equal to another number.

Soccer Connection

Coaches sometimes plan how much of a match should focus on attacking or defending. Fractions can help show how much time a team spends using each strategy during a game.

Did You Know?

Xavi Hernández, one of the greatest midfielders in soccer history, was famous for controlling possession of the ball. His teams often kept the ball for long periods, sometimes more than **60% of the match**.

Worked Examples

Example 1

A team spends **2/4 of the match attacking**.

This means half the match is spent attacking.

2/4 = 1/2

Example 2

A team spends **3/4 of the match controlling possession**.

3/4 is more than 1/2, so the team controlled the ball for most of the game.

Example 3

A coach divides the field into **4 equal sections**.

A drill uses **3 of the sections**.

This fraction is:

3/4

Practice Problems

1. A team spends **2/3 of the match attacking** and **1/3 defending**.
   Which fraction is larger?

2. At **Camp Nou**, a team controls the ball for **3/5 of the match** and the opponent controls it for **2/5**.
   Which fraction is larger?

3. A coach divides the field into **4 equal sections**, and a drill uses **3 sections**.
   What fraction of the field is used?

4. **Lionel Messi** takes **2 of 6 shots on goal** for his team.
   What fraction of the shots did he take?

5. A team spends **4 of 8 practice drills working on defense**.
   What fraction of practice is spent on defense?

6. At **Allianz Arena**, a team controls the ball for **5/10 of the match**.
   What fraction of the match is that?

7. A coach divides the field into **6 equal sections**, and a drill uses **4 sections**.
   What fraction of the field is used?

8. At **Old Trafford**, a team scores **3 goals out of 6 shots**.
   What fraction of shots were goals?

9. A team spends **2 of 4 practice drills working on shooting**.
   What fraction of practice is spent on shooting?

10. At **Estadio Azteca**, a team controls the ball for **7/10 of the match**.
    What fraction of the match is controlled?

Answers

1. 2/3 | 2. 3/5 | 3. 3/4 | 4. 2/6 | 5. 4/8 | 6. 5/10 | 7. 4/6 | 8. 3/6 | 9. 2/4 | 10. 7/10

Challenge Problem

At **Maracanã Stadium**, a team divides practice time into **8 equal parts**.

- **3 parts are spent practicing passing**
- **2 parts are spent practicing shooting**
- **3 parts are spent practicing defense**

What fraction of practice is spent on **passing and defense combined**?

3/8 + 3/8 = **6/8**

Answer: **6/8 of practice** is spent on passing and defense.

29

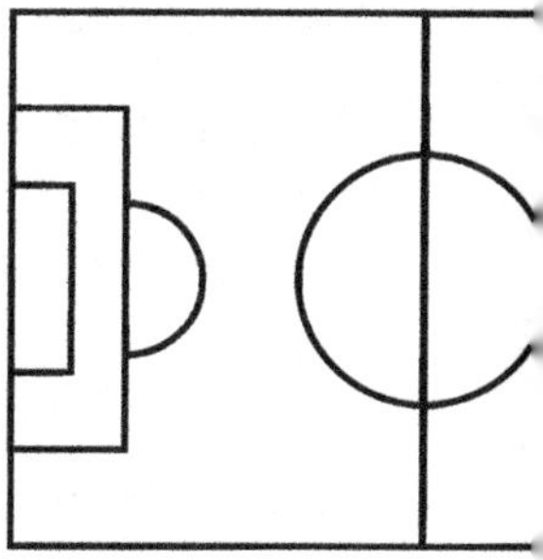

Review of Fractions

Common Core Standards: 3.NF.A.1, 3.NF.A.2, 3.NF.A.3

Introduction

Fractions help us understand **parts of a whole**. In soccer, fractions can describe parts of the field, parts of practice drills, or parts of a match. In this chapter, we will review the important ideas about fractions that you have learned so far.

Key Definitions

Fraction: A number that represents part of a whole.

Numerator: The top number of a fraction that shows how many parts we have.

Denominator: The bottom number of a fraction that shows how many equal parts the whole is divided into.

Soccer Connection

During practice sessions, coaches divide the soccer field, practice drills, and training time into equal parts. Fractions help describe how much of the field or practice time is used for each activity.

Did You Know?

The **FIFA Women's World Cup** is one of the fastest-growing sporting events in the world. Millions of fans watch the tournament as the best women's teams compete for the championship.

Worked Examples

Example 1

A soccer field is divided into **2 equal sections**.

Each section is:

1/2 of the field.

Example 2

A coach divides the field into **4 equal sections**, and a drill uses **3 sections**.

The fraction of the field used is:

3/4

Example 3

A practice session has **6 drills**, and **2 drills focus on shooting**.

The fraction of practice spent on shooting is:

2/6

Practice Problems

1. A soccer field is divided into **3 equal parts**.
   What fraction of the field is one part?
2. At **Camp Nou**, the field is divided into **4 equal sections**.
   What fraction of the field is one section?
3. A drill uses **2 of the 5 equal sections** of the field.
   What fraction of the field is used?
4. **Alex Morgan** leads a practice where **3 of 6 drills focus on passing**.
   What fraction of practice is spent on passing?
5. A team spends **4 of 8 drills practicing shooting**.
   What fraction of practice is spent on shooting?
6. At **Allianz Arena**, a team uses **5 of 10 sections of the field**.
   What fraction of the field is used?
7. A coach divides practice into **7 drills**, and **3 drills focus on defense**.
   What fraction of practice is spent on defense?
8. At **Old Trafford**, a team practices on **6 of 12 sections of the field**.
   What fraction of the field is used?
9. A team spends **2 of 4 drills practicing passing**.
   What fraction of practice is spent on passing?
10. At **Estadio Azteca**, a drill uses **7 of 8 sections of the field**.
    What fraction of the field is used?

Answers

1. 1/3 | 2. 1/4 | 3. 2/5 | 4. 3/6 | 5. 4/8 | 6. 5/10 | 7. 3/7 | 8. 6/12 | 9. 2/4 | 10. 7/8

Challenge Problem

At **Maracanã Stadium**, a coach divides the soccer field into **8 equal sections** for training.

- **3 sections are used for passing drills**
- **2 sections are used for shooting drills**
- **3 sections are used for defending drills**

What fraction of the field is used for **passing and shooting combined**?

3/8 + 2/8 = **5/8**

Answer: **5/8 of the field**

30

Quarter 3 Challenge Cup

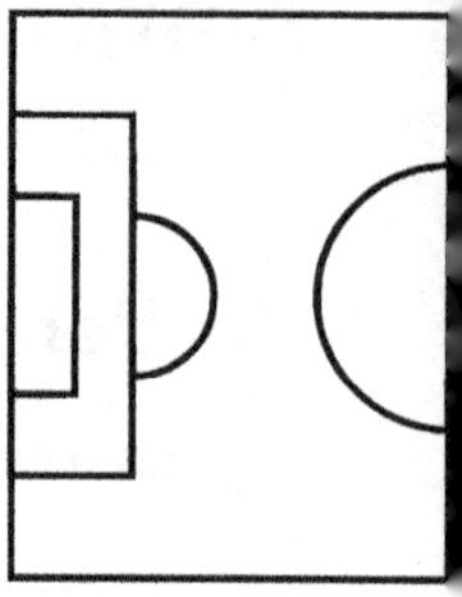

Common Core Standards: 3.NF.A.1, 3.NF.A.2, 3.NF.A.3

Introduction

In this part of the book, we learned how fractions represent **parts of a whole**. Fractions can describe parts of the soccer field, parts of practice drills, or parts of a match. In this chapter, you will review and apply what you have learned about fractions.

Key Definitions

Fraction: A number that represents part of a whole.

Numerator: The top number of a fraction showing how many parts are being counted.

Denominator: The bottom number showing how many equal parts the whole is divided into.

Soccer Connection

Soccer practices often divide the field and training time into sections so that multiple drills can happen at once. Fractions help coaches and players understand how much of the field or practice is being used.

Did You Know?

The **UEFA Women's Champions League** is the top club competition in European women's soccer. The best teams from many countries compete each year for the championship.

Worked Examples

Example 1

A coach divides the field into **4 equal sections**.

Each section represents:

1/4 of the field.

Example 2

A drill uses **3 of the 4 sections** of the field.

The fraction of the field used is:

3/4

Example 3

A practice session has **6 drills**, and **2 drills focus on shooting**.

The fraction of practice spent on shooting is:

2/6

Practice Problems

1. A soccer field is divided into **2 equal sections**.
   What fraction of the field is one section?

2. At **Camp Nou**, a drill uses **2 of the 4 sections** of the field.
   What fraction of the field is used?

3. A coach divides the field into **5 equal sections**.
   What fraction of the field is one section?

4. **Lionel Messi** practices shooting in **3 of 6 drills**.
   What fraction of practice is spent on shooting?

5. A team uses **4 of 8 sections** of the field for practice.
   What fraction of the field is used?

6. At **Allianz Arena**, a team practices on **3 of 5 sections** of the field.
   What fraction of the field is used?

7. A coach divides practice into **7 drills**, and **2 drills focus on passing**.
   What fraction of practice is spent on passing?

8. At **Old Trafford**, a drill uses **6 of 10 sections** of the field.
   What fraction of the field is used?

9. A team spends **5 of 8 drills practicing defense**.
   What fraction of practice is spent on defense?

10. At **Estadio Azteca, 3 of 4 sections of the field** are used for a scrimmage.
    What fraction of the field is used?

Answers

1. 1/2	2. 2/4	3. 1/5	4. 3/6	5. 4/8	6. 3/5	7. 2/7	8. 6/10	9. 5/8	10. 3/4

Challenge Problem

At **Maracanã Stadium**, the soccer field is divided into **10 equal sections** for a practice session.

- **4 sections are used for passing drills**
- **3 sections are used for shooting drills**

What fraction of the field is used for **passing and shooting combined**?

4/10 + 3/10 = **7/10**

Answer: **7/10 of the field**

31

Rounding Numbers in Soccer Statistics

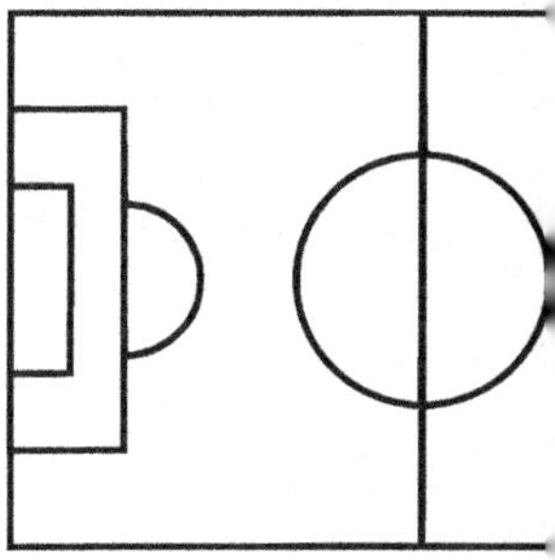

Common Core Standard: 3.NBT.A.1

Introduction

In soccer statistics, numbers can sometimes be large and difficult to work with. **Rounding numbers** helps make numbers easier to understand and compare. When we round a number, we replace it with the nearest **ten or hundred**.

Key Definitions

Round: To change a number to the nearest ten or hundred.

Place Value: The value of a digit depending on its position in a number.

Soccer Connection

Coaches and analysts often look at statistics such as passes completed, distance run, or number of fans attending a match. Rounding these numbers can make it easier to understand large statistics quickly.

Did You Know?

The stadium **Camp Nou** in Barcelona has one of the largest seating capacities in soccer, with around **99,000 seats** for fans.

Worked Examples

Example 1

A player completes **47 passes** in a match.

Rounded to the nearest ten:

$47 \rightarrow$ **50**

Example 2

At **Wembley Stadium**, a team completes **63 passes** during a drill.

Rounded to the nearest ten:

63 → **60**

Example 3

A goalkeeper makes **88 saves during a season**.

Rounded to the nearest ten:

88 → **90**

Practice Problems

1. A player completes **34 passes** in a match.
   Round to the nearest ten.
2. At **Camp Nou**, a team makes **67 passes** during a drill.
   Round to the nearest ten.
3. A goalkeeper makes **52 saves** during a season.
   Round to the nearest ten.
4. **Lionel Messi** attempts **76 shots** during a season.
   Round to the nearest ten.
5. A team completes **43 corner kicks** in a season.
   Round to the nearest ten.
6. At **Allianz Arena**, a team attempts **91 shots** during a tournament.
   Round to the nearest ten.
7. A player completes **28 tackles** during the season.
   Round to the nearest ten.
8. At **Old Trafford**, a team scores **84 goals** over several matches.
   Round to the nearest ten.
9. A goalkeeper makes **65 saves** during a tournament.
   Round to the nearest ten.
10. At **Estadio Azteca**, a team completes **72 passes** during a drill.
    Round to the nearest ten.

Answers

1. 30	2. 70	3. 50	4. 80	5. 40	6. 90	7. 30	8. 80	9. 70	10. 70

Challenge Problem

A team completes **48 passes in the first half** and **57 passes in the second half**.

1. Round each number to the nearest ten.
2. What is the total number of passes after rounding?

$48 \rightarrow 50$

$57 \rightarrow 60$

$50 + 60 =$ **110**

Answer: **110 passes (rounded)**

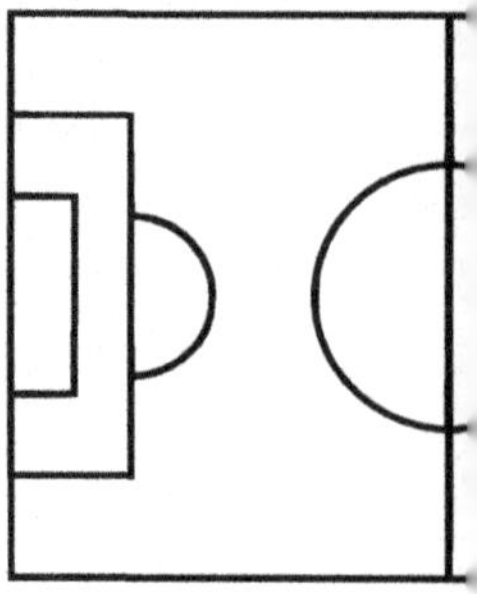

32

Adding and Subtracting Soccer Scores

Common Core Standard: 3.NBT.A.2

Introduction

Soccer teams often play many matches during a season or tournament. To understand a team's performance, we sometimes need to **add or subtract scores** from different games. Learning to add and subtract larger numbers helps us track goals scored and goals allowed over time.

Key Definitions

Sum: The answer when numbers are added together.

Difference: The answer when one number is subtracted from another.

Soccer Connection

Teams often track statistics such as **total goals scored** or **goal difference**. Goal difference is the number of goals a team scores minus the number of goals it allows.

Did You Know?

Some soccer matches can end with many goals scored, while others may end with only one or two goals. Teams often analyze scoring statistics to improve their performance.

Worked Examples

Example 1

A team scores **12 goals in one tournament** and **15 goals in another**.

12 + 15 = 27

Total goals scored = **27**

Example 2

At **Wembley Stadium**, a team scores **18 goals** during the season but allows **11 goals**.

18 − 11 = 7

Goal difference = **7**

Example 3

A player scores **14 goals in the first half of the season** and **9 goals in the second half.**

14 + 9 = 23

Total goals scored = **23**

Practice Problems

1. A team scores **13 goals in one tournament** and **16 goals in another.**
   How many goals did the team score in total?
2. At **Camp Nou**, a team scores **20 goals** but allows **12 goals.**
   What is the goal difference?
3. A player scores **11 goals in the first half of the season** and **14 goals in the second half.**
   How many goals did the player score in total?
4. **Cristiano Ronaldo** scores **18 goals in league matches** and **7 goals in cup matches.**
   How many goals did he score in total?
5. A team scores **25 goals during a tournament** but allows **19 goals.**
   What is the goal difference?
6. At **Allianz Arena**, a team scores **14 goals in the first half of the season** and **17 goals in the second half.**
   How many goals were scored in total?
7. A player scores **9 goals in home games** and **8 goals in away games.**
   How many goals did the player score?
8. At **Old Trafford**, a team scores **22 goals** but allows **15 goals.**
   What is the goal difference?
9. A team scores **16 goals in one tournament** and **13 goals in another.**
   How many goals were scored in total?
10. At **Estadio Azteca**, a team scores **21 goals** but allows **17 goals.**
    What is the goal difference?

Answers

1. 29	2. 8	3. 25	4. 25	5. 6	6. 31	7. 17	8. 7	9. 29	10. 4

Challenge Problem

During a season at **Maracanã Stadium**, a team scores **18 goals in the first half of the season** and **22 goals in the second half**. The team allows **25 goals** during the season.

1. How many total goals did the team score?
2. What is the team's goal difference?

Step 1

$18 + 22 = 40$

Step 2

$40 - 25 = 15$

Answers

1. 40 goals scored
2. Goal difference = 15

33

Measuring Soccer Field Distances

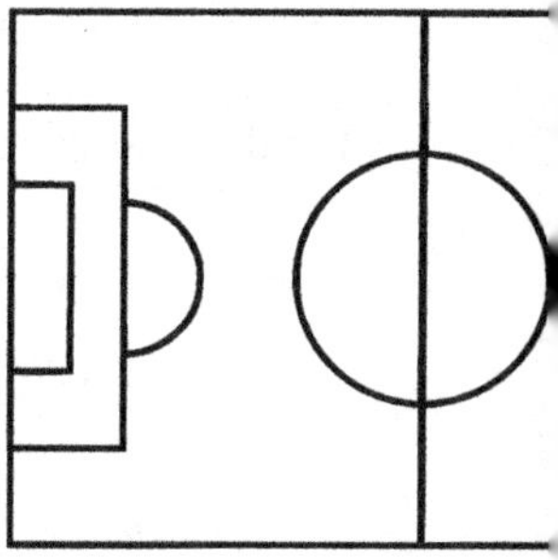

Common Core Standard: 3.MD.A.2

Introduction

Measurement helps us understand **distance, length, and size**. In soccer, measurement is important for understanding the size of the field, the distance of passes, and the length of runs during a match. Players and coaches often measure distances in meters.

Key Definitions

Length: The measurement of how long something is.

Meter: A unit used to measure distance.

Soccer Connection

Soccer fields have standard measurements. Players also measure distances during training drills, such as how far they pass the ball or how far they run during conditioning exercises.

Did You Know?

A professional soccer field is usually about **100–110 meters long** and **64–75 meters wide**, depending on the stadium and competition rules.

Worked Examples

Example 1

A player passes the ball **12 meters**, then passes it **8 meters**.

12 + 8 = 20

Total distance of passes = **20 meters**

Example 2

At **Camp Nou**, a player runs **15 meters**, then runs another **10 meters**.

15 + 10 = 25

Total distance run = **25 meters**

Example 3

A pass travels **30 meters**, but the receiver moves **5 meters back** to control the ball.

30 − 5 = 25

Distance from starting point = **25 meters**

Practice Problems

1. A player runs **10 meters** and then runs **15 meters**.
   What is the total distance run?

2. At **Wembley Stadium**, a pass travels **18 meters**, and another pass travels **12 meters**.
   What is the total distance of the passes?

3. A player runs **20 meters** forward and **5 meters back**.
   What is the total forward distance from the starting point?

4. **Lionel Messi** dribbles the ball **14 meters** and then passes it **11 meters**.
   What is the total distance the ball traveled?

5. A player runs **16 meters**, then runs **9 meters more**.
   What is the total distance run?

6. At **Allianz Arena**, a pass travels **22 meters**, then another pass travels **13 meters**.
   What is the total distance?

7. A player runs **25 meters** and then runs **7 meters back**.
   What is the total forward distance from the starting point?

8. At **Old Trafford**, a pass travels **19 meters**, followed by another pass of **10 meters**.
   What is the total distance of the passes?

9. A player dribbles **13 meters** and then runs **12 meters**.
   What is the total distance?

10. At **Estadio Azteca**, a player runs **17 meters**, then runs **8 meters more**.
    What is the total distance run?

Answers

1. 25 meters	2. 30 meters	3. 15 meters	4. 25 meters	5. 25 meters

6. 35 meters	7. 18 meters	8. 29 meters	9. 25 meters	10. 25 meters

Challenge Problem

At **Maracanã Stadium**, a player runs **20 meters**, then passes the ball **18 meters**, and then runs **12 meters** to receive the ball again.

What is the total distance the player and ball traveled?

20 + 18 + 12 = **50 meters**

Answer: **50 meters**

34

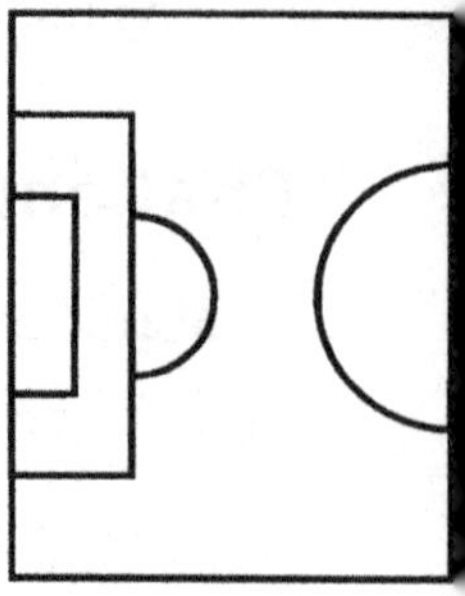

Time in a Soccer Match

Common Core Standard: 3.MD.A.1

Introduction

Time is very important in soccer. A professional match lasts **90 minutes**, divided into two halves of **45 minutes each**. Understanding how to add and subtract time helps players, coaches, and fans keep track of the game.

Key Definitions

Minute: A unit used to measure time.

Elapsed Time: The amount of time that has passed from the start of an event.

Soccer Connection

During a match, coaches track how much time has passed and how much time remains. They may change strategies depending on whether the match is just beginning, halfway finished, or close to the end.

Did You Know?

In addition to the 90 minutes of regular play, referees may add **stoppage time** to make up for delays during the match. This extra time is sometimes called **injury time**.

Worked Examples

Example 1

A soccer match begins at **3:00 PM**.
Halftime begins **45 minutes later**.

3:00 + 45 minutes = **3:45 PM**

Example 2

The second half begins at **4:00 PM**.
After **30 minutes**, the time is:

4:00 + 30 minutes = **4:30 PM**

Example 3

A match starts at **2:00 PM** and ends **90 minutes later**.

2:00 + 90 minutes = **3:30 PM**

Practice Problems

1. A soccer match starts at **1:00 PM**.
   Halftime is **45 minutes later.**
   What time does halftime begin?

2. At **Wembley Stadium**, the second half begins at **4:15 PM**.
   What time will it be **30 minutes later**?

3. A match begins at **2:30 PM** and lasts **90 minutes**.
   What time does the match end?

4. **Cristiano Ronaldo** scores a goal **20 minutes** after a match begins at **3:00 PM**.
   What time is the goal scored?

5. At **Camp Nou**, a match begins at **5:00 PM**.
   Halftime begins **45 minutes later**.
   What time is halftime?

6. A match begins at **6:30 PM**.
   After **30 minutes**, what time is it?

7. At **Allianz Arena**, a match begins at **7:15 PM**.
   Halftime is **45 minutes later.**
   What time is halftime?

8. A match begins at **2:45 PM** and lasts **90 minutes**.
   What time does the match end?

9. At **Old Trafford**, the second half begins at **8:00 PM**.
   What time will it be **25 minutes later**?

10. At **Estadio Azteca**, a match begins at **1:30 PM**.
    After **45 minutes**, what time is halftime?

Answers

1. 1:45 PM	2. 4:45 PM	3. 4:00 PM	4. 3:20 PM	5. 5:45 PM

6. 7:00 PM	7. 8:00 PM	8. 4:15 PM	9. 8:25 PM	10. 2:15 PM

Challenge Problem

A soccer match begins at **3:15 PM**.

- Halftime occurs **45 minutes later**
- The second half lasts another **45 minutes**

What time does the match end?

Step 1

3:15 + 45 minutes = **4:00 PM**

Step 2

4:00 + 45 minutes = **4:45 PM**

Answer: **4:45 PM**

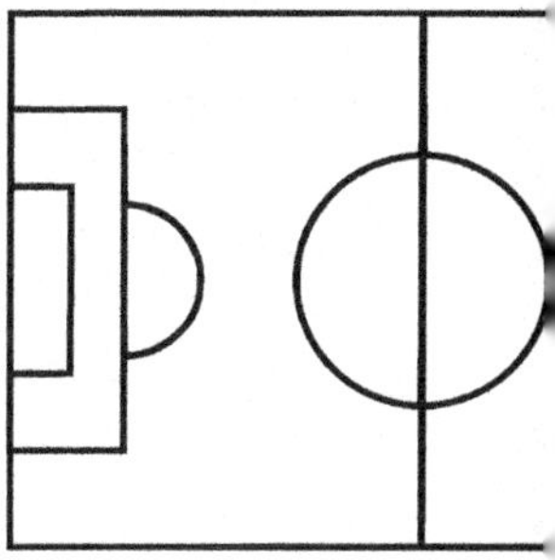

35

Area of a Soccer Practice Grid

Common Core Standard: 3.MD.C.5–7

Introduction

Area tells us how much **space a shape covers**. In soccer training, coaches often set up practice grids using cones. Understanding area helps players and coaches know how much space a drill uses on the field.

Key Definitions

Area: The amount of space inside a shape.

Square Unit: A unit used to measure area.

Soccer Connection

Coaches frequently use square practice grids for drills such as passing, dribbling, and small-sided games. The area of the grid helps determine how much space players have to move.

Did You Know?

Small-sided soccer games, such as **4 vs 4**, are often played in smaller practice grids. These games help players improve passing, dribbling, and quick decision-making.

Worked Examples

Example 1

A practice grid has **4 rows of 5 squares**.

$4 \times 5 = 20$

Area = **20 square units**

Example 2

At **Camp Nou**, a drill grid has **6 rows of 3 squares**.

$6 \times 3 = 18$

Area = **18 square units**

Example 3

A grid has **5 rows of 5 squares**.

$5 \times 5 = 25$

Area = **25 square units**

Practice Problems

1. A practice grid has **3 rows of 4 squares**.
   What is the area?
2. At **Wembley Stadium**, a grid has **5 rows of 4 squares**.
   What is the area?
3. A drill grid has **6 rows of 2 squares**.
   What is the area?
4. **Lionel Messi** practices in a grid with **4 rows of 6 squares**.
   What is the area?
5. A coach sets up a grid with **7 rows of 3 squares**.
   What is the area?
6. At **Allianz Arena**, a grid has **5 rows of 5 squares**.
   What is the area?
7. A drill grid has **8 rows of 2 squares**.
   What is the area?
8. At **Old Trafford**, a grid has **4 rows of 7 squares**.
   What is the area?
9. A coach sets up a grid with **3 rows of 9 squares**.
   What is the area?
10. At **Estadio Azteca**, a grid has **6 rows of 4 squares**.
   What is the area?

Answers

1. 12 square units	2. 20 square units	3. 12 square units	4. 24 square units	5. 21 square units

6. 25 square units	7. 16 square units	8. 28 square units	9. 27 square units	10. 24 square units

Challenge Problem

At **Maracanã Stadium**, a coach sets up a large training grid with **8 rows of 6 squares**. Later, another grid with **4 rows of 6 squares** is added next to it.

What is the **total area** of both grids combined?

Step 1

$8 \times 6 = 48$

Step 2

$4 \times 6 = 24$

Step 3

$48 + 24 =$ **72**

Answer: **72 square units**

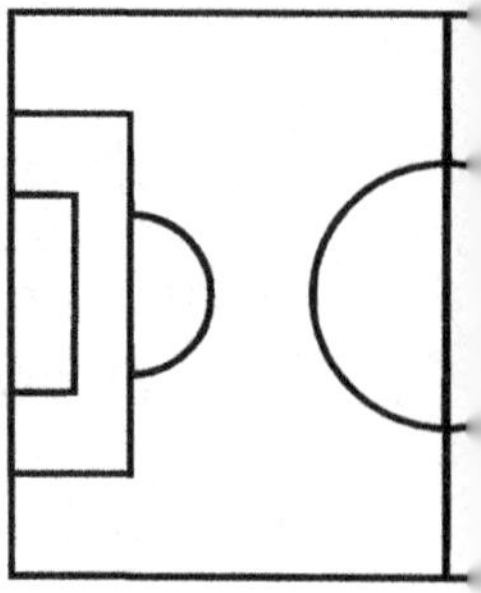

36

Perimeter of a Soccer Field

Common Core Standard: 3.MD.D.8

Introduction

The **perimeter** of a shape is the total distance around it. In soccer, understanding perimeter helps us measure the boundaries of practice fields, training grids, or parts of the soccer field.

Key Definitions

Perimeter: The total distance around a shape.

Length: The measurement of how long something is from one end to the other. It is usually the **longest side** of an object.

Width: The measurement of how wide something is from side to side. It is usually the **shorter side**, measured across the object.

Soccer Connection

Coaches often set up rectangular training areas with cones. Knowing the perimeter helps determine how much boundary space is needed for a drill or small-sided game.

Did You Know?

Professional soccer fields vary in size, but many are about **105 meters long and 68 meters wide** in international matches.

Worked Examples

Example 1

A practice field is **10 meters long** and **6 meters wide**.

Perimeter = 10 + 10 + 6 + 6

Perimeter = **32 meters**

Example 2

At **Wembley Stadium**, a training grid is **12 meters long** and **8 meters wide**.

Perimeter = 12 + 12 + 8 + 8

Perimeter = **40 meters**

Example 3

A drill area is **9 meters long** and **5 meters wide**.

Perimeter = 9 + 9 + 5 + 5

Perimeter = **28 meters**

Practice Problems

1. A training grid is **8 meters long** and **6 meters wide**.
   What is the perimeter?
2. At **Camp Nou**, a drill field is **10 meters long** and **7 meters wide**.
   What is the perimeter?
3. A practice area is **9 meters long** and **4 meters wide**.
   What is the perimeter?
4. **Cristiano Ronaldo** trains in a grid that is **12 meters long** and **5 meters wide**.
   What is the perimeter?
5. A coach sets up a field that is **7 meters long** and **7 meters wide**.
   What is the perimeter?
6. At **Allianz Arena**, a drill area is **14 meters long** and **6 meters wide**.
   What is the perimeter?
7. A training grid is **11 meters long** and **3 meters wide**.
   What is the perimeter?
8. At **Old Trafford**, a field is **8 meters long** and **8 meters wide**.
   What is the perimeter?
9. A drill area is **15 meters long** and **5 meters wide**.
   What is the perimeter?
10. At **Estadio Azteca**, a grid is **9 meters long** and **6 meters wide**.
    What is the perimeter?

Answers

1. 28 meters	2. 34 meters	3. 26 meters	4. 34 meters	5. 28 meters
6. 40 meters	7. 28 meters	8. 32 meters	9. 40 meters	10. 30 meters

Challenge Problem

At **Maracanã Stadium**, a training field is **20 meters long** and **12 meters wide**. A coach runs **two laps around the field** during warmups.

Step 1

Perimeter = 20 + 20 + 12 + 12

Perimeter = 64 meters

Step 2

Two laps = 64 × 2

Total distance = **128 meters**

Answer: **128 meters**

37

Bar Graphs of Goals Scored

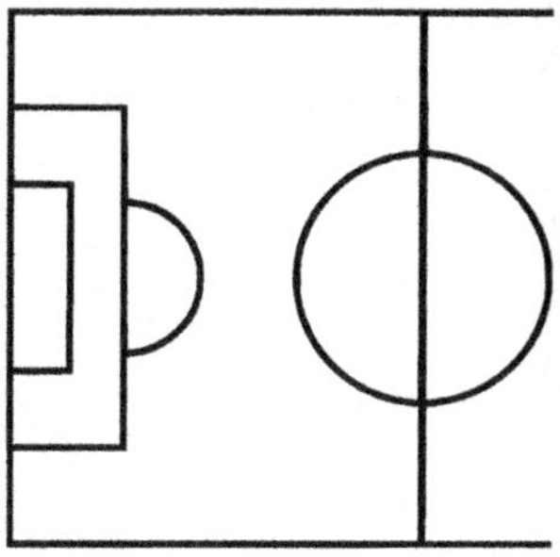

Common Core Standard: 3.MD.B.3

Introduction

Graphs help us organize and understand information. A **bar graph** uses bars to show and compare numbers. In soccer, bar graphs can help us compare how many goals players or teams have scored.

Key Definitions

Bar Graph: A chart that uses bars to represent numbers.

Data: Information that can be collected and analyzed.

Soccer Connection

Coaches and analysts often track statistics such as goals scored, shots taken, or saves made. Bar graphs make it easier to compare these numbers quickly.

Did You Know?

Many professional teams now use **data analysis** to evaluate player performance. Analysts study statistics such as goals, assists, passes, and distance run during matches.

Worked Examples

Example 1

A bar graph shows goals scored by three players:

Player A: 4 goals
Player B: 6 goals
Player C: 3 goals

The player with the most goals is **Player B**.

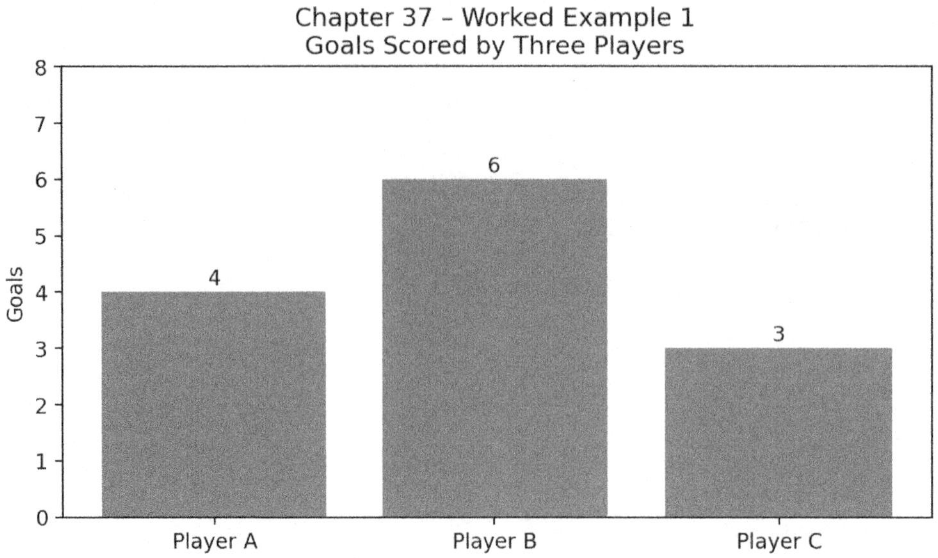

Example 2

At **Camp Nou**, a graph shows goals scored in four matches:

Match 1: 2 goals

Match 2: 3 goals

Match 3: 1 goal

Match 4: 4 goals

The most goals were scored in **Match 4**.

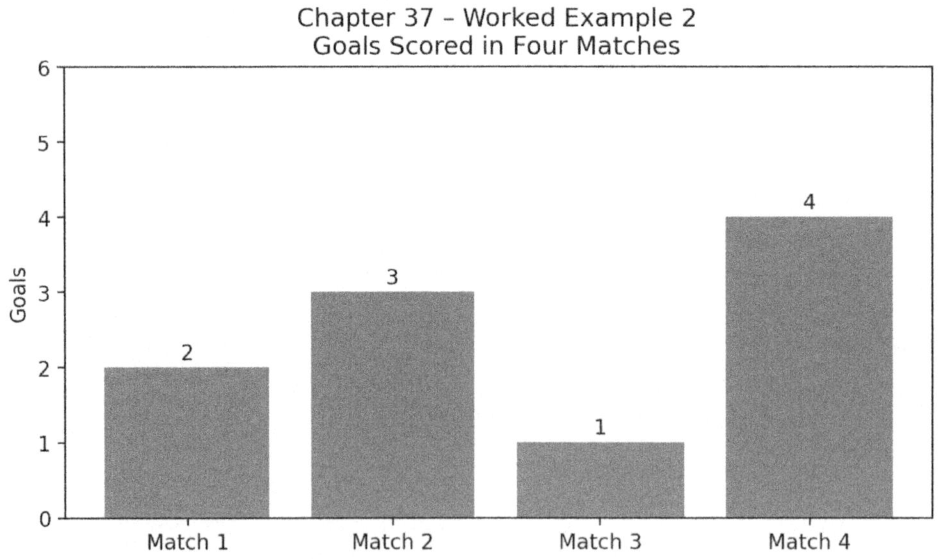

Example 3

A team's bar graph shows goals scored by three players:

Messi: 5

Mbappé: 3

Haaland: 4

The player with the most goals is **Messi**.

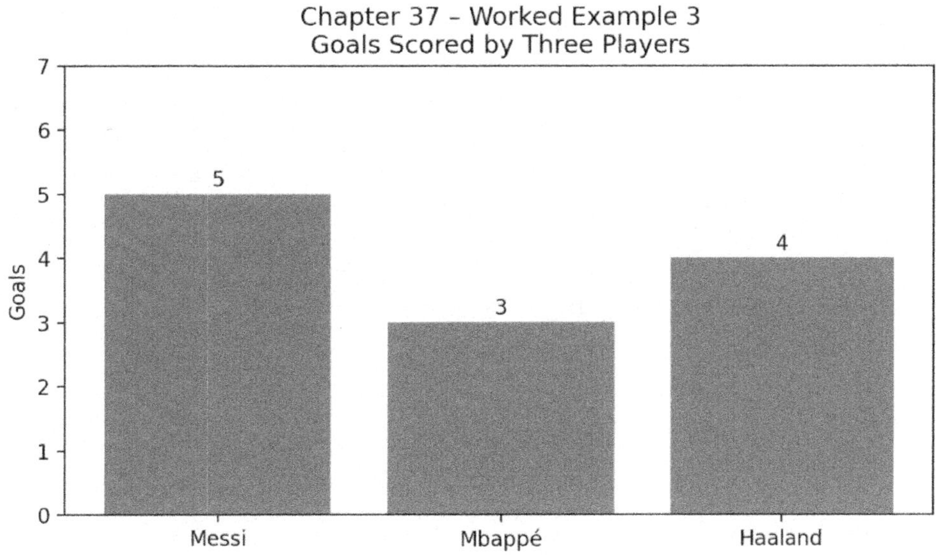

Practice Problems

A team keeps track of goals scored in five matches.

Match 1: 2 goals

Match 2: 4 goals

Match 3: 3 goals

Match 4: 5 goals

Match 5: 1 goal

1. Which match had the most goals?
2. How many goals were scored in Match 3?
3. How many goals were scored in Match 5?
4. What is the total number of goals scored in Match 1 and Match 2?
5. What is the total number of goals scored in Match 3 and Match 4?
6. At **Wembley Stadium**, which match had the fewest goals?
7. How many goals were scored in Match 4?
8. What is the difference between goals in Match 2 and Match 5?
9. How many goals were scored in the first three matches combined?
10. What is the total number of goals scored in all five matches?

Answers

1. Match 4	2. 3	3. 1	4. 6	5. 8

6. Match 5	7. 5	8. 3	9. 9	10. 15

Challenge Problem

At **Old Trafford**, a team tracks goals scored in four matches.

Match 1: 4 goals
Match 2: 3 goals
Match 3: 5 goals
Match 4: 4 goals

What is the **average number of goals per match**?

Step 1

$4 + 3 + 5 + 4 = 16$

Step 2

$16 \div 4 = \mathbf{4}$

Answer: **4 goals per match (average)**

38

Line Plots of Player Statistics

Common Core Standard: 3.MD.B.4

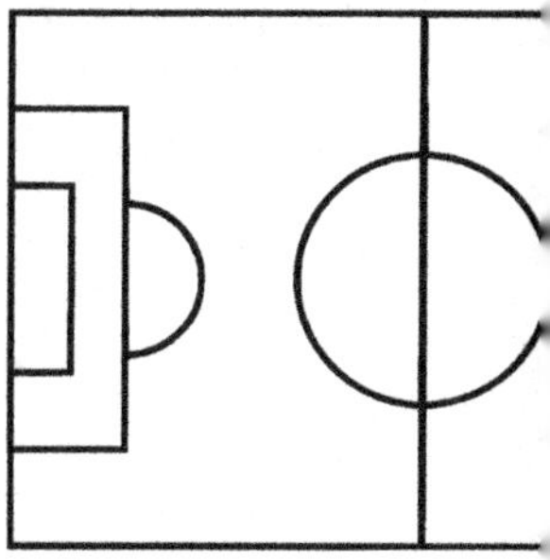

Introduction

A **line plot** is another way to display data. It shows numbers along a number line and marks how often each value appears. Line plots are useful for understanding patterns in statistics such as goals, passes, or saves during soccer matches.

Key Definitions

Line Plot: A graph that shows data points above a number line.

Data: Information collected and organized for study.

Soccer Connection

Teams track many statistics during matches and practices. Line plots can help coaches quickly see patterns, such as how many goals were scored in different games or how many saves a goalkeeper made.

Did You Know?

Modern soccer teams collect thousands of data points during a match, including passes, runs, shots, and tackles. Analysts study this information to help teams improve their performance.

Worked Examples

Example 1

Goals scored in five matches:

2, 3, 2, 4, 3

A line plot would show:

2 appearing twice
3 appearing twice

4 appearing once

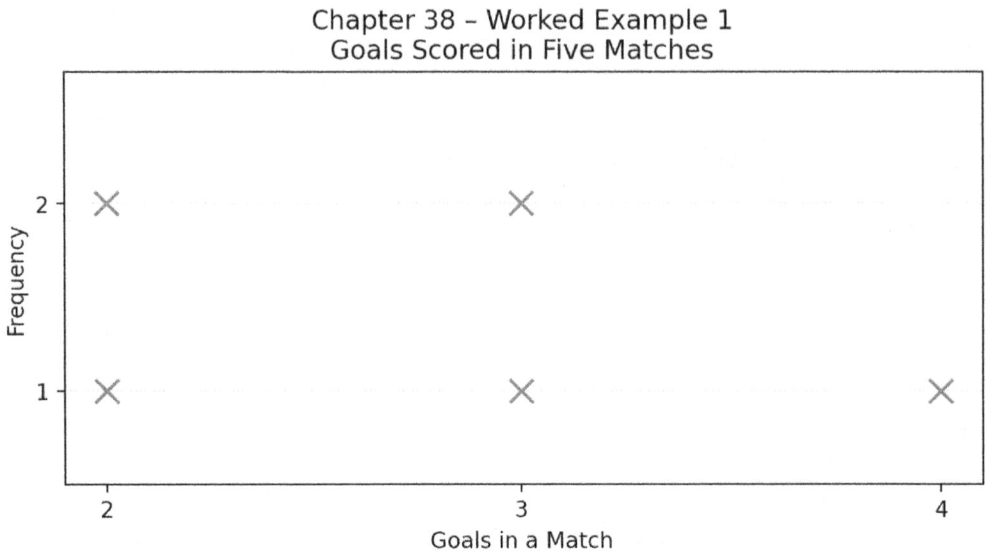

Example 2

Saves made by a goalkeeper in five games:

3, 5, 4, 5, 2

The number **5** appears most often.

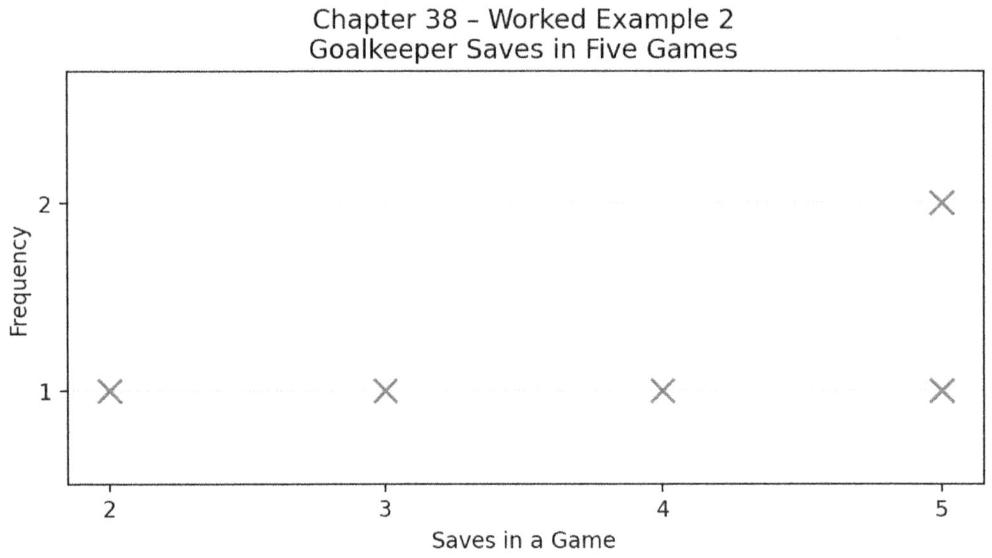

Example 3

Passes completed during short drills:

6, 7, 7, 8, 6

The values **6** and **7** appear twice.

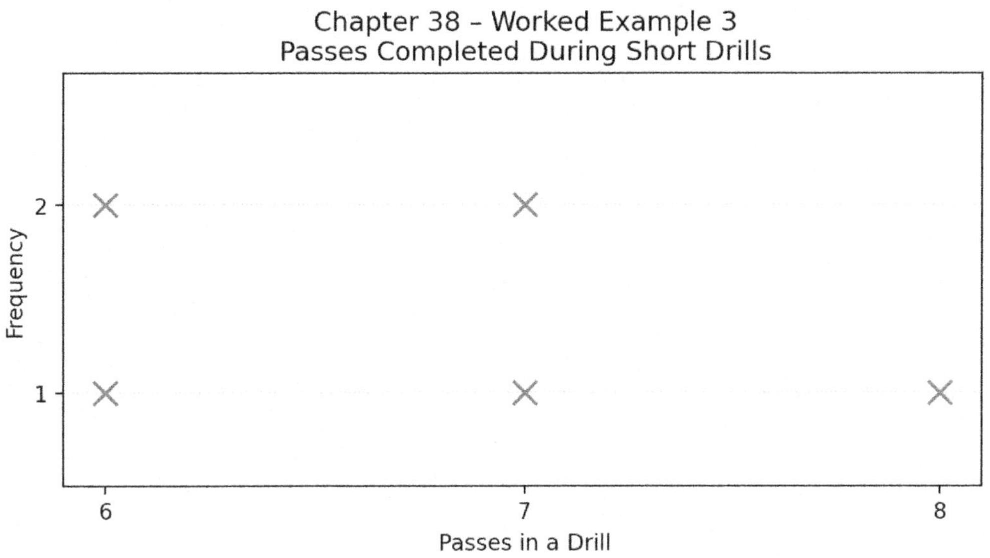

Practice Problems

A team records the number of goals scored in **six matches**:

2, 3, 3, 4, 2, 5

1. How many matches had **2 goals**?
2. How many matches had **3 goals**?
3. How many matches had **4 goals**?
4. How many matches had **5 goals**?
5. Which number of goals happened **most often**?
6. At **Camp Nou**, how many total matches are represented in the data?
7. How many matches had **3 or more goals**?
8. What is the **largest number of goals** scored in a match?
9. What is the **smallest number of goals** scored in a match?
10. What is the **total number of goals scored across all matches**?

Answers

1. 2	2. 2	3. 1	4. 1	5. 2 goals and 3 goals (tie)
6. 6 matches	7. 4 matches	8. 5	9. 2	10. 19

Challenge Problem

At **Wembley Stadium**, a team records goals scored in **seven matches**:

1, 2, 2, 3, 3, 3, 4

1. Which number of goals appears **most often**?

2. How many matches had **3 goals**?

Answers:

1. 3 goals

2. 3 matches

39

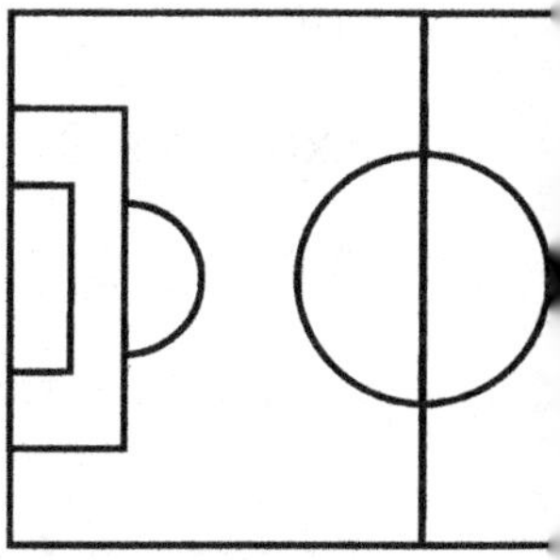

Shapes on the Soccer Field

Common Core Standard: 3.G.A.1

Introduction

Geometry is the study of **shapes and their properties**. Many shapes appear on a soccer field, including rectangles, circles, and squares. Understanding shapes helps us describe different parts of the field and training areas.

Key Definitions

Rectangle: A shape with four sides and four right angles.

Square: A rectangle where all four sides are the same length.

Circle: A round shape where all points are the same distance from the center.

Soccer Connection

A soccer field contains many geometric shapes. The field itself is a **rectangle**, the **center circle** is a circle, and practice grids used for drills are often squares or rectangles.

Did You Know?

The **center circle** on a soccer field has a radius of **9.15 meters**. This space keeps opposing players a safe distance away during kickoff.

Worked Examples

Example 1

The soccer field is shaped like a **rectangle**.

A rectangle has **four sides** and **four right angles**.

Example 2

The **center circle** on the field is shaped like a **circle**.

All points on the edge of the circle are the same distance from the center.

Example 3

A practice grid with equal sides is a **square**.

A square has **four equal sides**.

Practice Problems

1. The shape of a soccer field is a ___________.
2. The center circle on a soccer field is shaped like a ___________.
3. A practice grid with **four equal sides** is called a ___________.
4. At **Camp Nou**, the penalty box forms part of a ___________ shape.
5. A shape with **four sides and four right angles** is called a ___________.
6. At **Wembley Stadium**, a practice grid with **four equal sides** is a ___________.
7. The center circle used during kickoff is a ___________.
8. A training field with **four sides but not all equal** is a ___________.
9. At **Allianz Arena**, the large playing area is shaped like a ___________.
10. A small practice grid where all sides are equal is a ___________.

Answers

1. Rectangle	2. Circle	3. Square	4. Rectangle	5. Rectangle
6. Square	7. Circle	8. Rectangle	9. Rectangle	10. Square

Challenge Problem

A coach sets up **three square practice grids** on the soccer field.

Each square has **4 equal sides**.

1. How many total sides do the three squares have?

Step 1

Each square has 4 sides.

Step 2

$3 \times 4 = 12$

Answer: **12 sides**

40

World Cup Math Challenge

Common Core Standards: Review of Grade 3 Standards

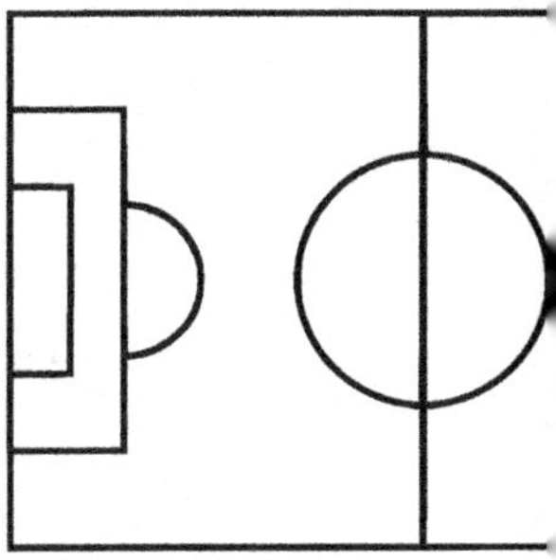

Introduction

Throughout this book, you have learned many math skills including multiplication, division, fractions, measurement, and geometry. In soccer, these skills help coaches organize teams, analyze statistics, and plan practices. In this final challenge chapter, you will apply everything you have learned to solve soccer-themed problems.

Key Definitions

Multiplication: A way to find the total number in equal groups.

Division: A way to split a total amount into equal groups.

Fraction: A number that represents part of a whole.

Soccer Connection

The **FIFA World Cup** is the largest soccer tournament in the world. Teams from many countries compete for the championship, and statistics from the tournament are analyzed by coaches, players, and fans.

Did You Know?

The **FIFA World Cup** takes place every **four years**, and teams from around the world compete through qualifying tournaments to earn a spot in the final competition.

Worked Examples

Example 1

A youth tournament has **4 teams with 8 players on each team**.

$4 \times 8 = 32$

Total players = **32**

Example 2

A team scores **3 goals in each of 5 matches.**

$3 \times 5 = 15$

Total goals = **15**

Example 3

A soccer field is divided into **4 equal sections**, and a drill uses **3 sections.**

The fraction of the field used is:

3/4

Practice Problems

1. A tournament has **5 teams with 7 players on each team.**
   How many players are in the tournament?

2. At **Wembley Stadium**, a team scores **2 goals in each of 6 matches.**
   How many goals are scored?

3. A coach divides **24 players** into **6 equal teams.**
   How many players are on each team?

4. **Lionel Messi** scores **3 goals in each of 4 games.**
   How many goals does he score?

5. A soccer field is divided into **4 equal sections**, and a drill uses **2 sections.**
   What fraction of the field is used?

6. A training grid has **6 rows of 4 squares.**
   What is the area?

7. A practice field is **10 meters long** and **5 meters wide.**
   What is the perimeter?

8. A team scores **12 goals** but allows **9 goals.**
   What is the goal difference?

9. A match begins at **2:30 PM** and lasts **90 minutes.**
   What time does the match end?

10. At **Camp Nou**, cones are arranged in **5 rows of 6 cones.**
    How many cones are used?

Answers

1. 35	2. 12	3. 4	4. 12	5. 2/4
6. 24 square units	7. 30 meters	8. 3	9. 4:00 PM	10. 30

Challenge Problem

At the **World Cup Final**, a youth soccer tournament has **6 teams with 8 players on each team.**

Step 1

$6 \times 8 = 48$ players

Later, the coach divides all the players into **4 equal groups for drills.**

Step 2

$48 \div 4 = 12$

Answer: **12 players per group**

Great job! You have completed the book.

Keep Practicing!

Math skills improve with practice, just like soccer skills improve with training. The more students practice solving problems, the more confident they will become.

Keep learning, keep practicing, and enjoy discovering how **math and soccer work together**!

Made in the USA
Monee, IL
07 November 2026

68083935R00077